FISHERY MANAGEMENT

FISHERY MANAGEMENT

By

Rekha R. Gaonkar

Maria D.C. Rodrigues

R. B. Patil

A.P.H. PUBLISHING CORPORATION
4435-36/7, ANSARI ROAD, DARYA GANJ
NEW DELHI-110 002

Published by
S.B. Nangia
A P H Publishing Corporation
7, Ansari Road, Daryaganj
New Delhi 110002
Ph.: 23274050
E-mail : aphbooks@gmail.com

2026

Printed at
Balaji Offset
Navin Shahdara, Delhi 110032

Dedicated to

Shri Vasantrao S. Joshi

Chairman, Murgaon Education Society, Vasco-da-Gama, Goa

For his constant support and encouragement in all academic endeavours

Vasantrao Joshi - A Crusader of Youth Empowerment

Shri Vasantrao Joshi, popularly known as Anna, the elder brother, is the finest flower of Goa's rich culture, ethos and tradition. Born in an aristocratic family of Vasco da Gama on 27th May 1921, he was educated in his native Vasco and then in Margao and Panaji in Marathi and Portuguese schools. While pursuing his Portuguese studies at the Govt. Lyceum of Panaji, at the tender age of sixteen Anna was drawn into his father's automobile business. Immediately after the end of Second World War, in 1946 he took full charge of this business, emerging as a major dealer of Bedford cars, trucks and buses. His firm, Auto service was Goa's first automobile importer of cars from England in specially chartered ships. Eventually, after the Liberation of Goa under Anna's leadership the House of Joshi's diversified into manufacturing of light engineering goods, Plastics and in Jute textiles, further adding new leaderships of two, three and four wheelers to their core business of automobiles.

Among the leading businessmen of Goa Anna stands very tall and high with his unmatched humanism and abiding concern for the underprivileged and boundless compassion for the downtrodden. For alleviating their plight he goes out all the way to do his best, bringing them a better lot and a comfortable life. The philosophy underlying his charity and philanthropy is that temporary relief and comfort are less important than empowerment of the individuals through education, skill formation and overall enlightenment. This rational explains the unstinted support and boundless patronage he has been extending to numerous educational institutions in Goa and outside. Education is his first love and prime concern. His aim is to provide the best of education even to the poorest child. For Anna best education is that which is value based and renders the learners self-confident, self-reliant imbued with lofty ideals as much as grit and determination to make a success of life.

An illustrious son of the soil, Anna has been repaying to the society at large its debt in manifold ways.

Well known connoisseur and patron of art, literature, theatre and education Anna had befriended many great personalities from various walks of culture like poet B. B. Borkar, P. L. Deshpande, Kusumagraj, Vasant Kanitkar and Prin T. K. Tope. Among the many important institutions with whose establishment and nurturing he has been associated, include Murgaon Hindu Samaj and Janata Vachanalaya of Vasco, Goa Education Society, Panaji, Shikshan Prasarak Mandali, Pune and Goa Hindu Association, Mumbai.

At the invitation of Shri D. B. Bandodkar, the first Chief Minister of Goa, Anna fought and won in 1972 the election as member of Goa Legislative Assembly. In fitness of things, the Assembly elected him as the Chairman of Public Accounts Committee. Goa Government appointed him also as the Chairman of the first Mormugao Planning and Development Authority, Goa Bagayatdar Co-operative Society etc.

Anna was instrumental in setting up of Vasco's first full fledged Co-operative Departmental Store, the Pushpagandha Co-operative Consumer Store, Ltd.

In the realm of higher education of Goa in general and Mormugao in particular Anna's richest contribution has been the establishment and nurturing from strength to strength the Murgaon Education Society which now accounts for educating annually over 2000 students through its 3 premier institutions, M.E.S. College of Arts & Commerce, M.E.S. Higher Secondary School of Arts, Science, Commerce and Vocational Studies and M.E.S. Institute of Management Studies and Research. Anna's enlightened and benevolent leadership and patronage are the key factors for the glorious progress of all these institutions. Even his occasional visits to the sprawling, evergreen campuses of these institutions radiate charm and inspire confidence among the students, teachers and staff whose hearts are always filled with deepest feelings of gratitude, respect and reverence for the towering personality, named Anna.

His much splendoured contribution to the enrichment of social, economic, political and cultural life of Goa immensely qualifies Anna to be hailed as one of the makers of modern Goa. Among the galaxy of great Goans he stands out as a staunch crusader of youth empowerment.

PREFACE

This book is the outcome of a UGC sponsored Major Research Project entitled, 'Fishing in Goa: An Inter-disciplinary Study of Its Socio-Economic and Political Aspects'. It was primarily an inter-disciplinary study conducted by the authors belonging to the disciplines of Economics, Sociology and Political Science. In recent years inter-disciplinary studies have assumed significance, as the correct comprehension of any issue requires multiple approach. Fisheries sector is one such issue, which is based on multiple factors like economic, social, political, environmental etc. Hence, this study was undertaken to understand the interplay of socio-economic and political factors in the fisheries sector.

The contemporary world is facing innumerable problems of varied nature. Many of these problems long exist and many more have emerged in recent years. To a great extent all these problems are man-made. The insatiable greed of man, ruthless exploitation of natural resources, mismanagement, corruption, lack of sincerity and commitment, insensitivity etc. are responsible for the rise of these problems. These problems are the main source of social suffering. All sections of the society are affected by this social suffering in one way or the other. The issues confronting the fisheries are one among these problems of the today's world.

Several problems have cropped up in course of fisheries development, which have culminated in the suffering of fisher folk, particularly the traditional fisher folk. The fisheries sector appears to be in crisis. The main focus of fisheries development was on the maximization of production to meet the food requirement of increasing population and to earn foreign- exchange. This emphasis on high production brought in new technology and huge finance in the fisheries sector. The process of modernization, no doubt resulted in unprecedented fish production (blue revolution) and increased incomes, but has taken a heavy toll. The accumulation of capital in the hands of few, over-exploitation of marine resources, depletion of precious marine resources, unemployment, replacement of old patterns of relations with individualistic and competition patterns and so on are the issues of concern, which emerged in the wake of the process of modernization.

The fisheries sector in Goa has not remained isolated from these processes of modernization. The fishing industry in Goa has experienced many structural changes due to the introduction of new and more efficient technology. There has been an unprecedented increase in the number of fleets leading to over exploitation of marine resources. The cliché "too many fishermen chasing too few fish" is aptly applicable in Goa as the number of mechanized boats is much higher than what 105 km of Goa's coastline can sustain. The new technology and consequent increase in fish production have brought changes in the structure of market and pricing patterns as well. These changes in fisheries sector have made adverse effects particularly on the traditional fishers as they didn't have sufficient resources to adapt to these changes. The new changes have posed a threat to sustainable development of fisheries in Goa.

The above issues confronting the fisheries sector have attracted the attention of academicians, statesmen, policymakers etc. These issues need to be addressed immediately and sincerely before the unrecoverable damage is caused. Being sensitive and close to these issues, they enkindled in us the idea of taking up a major research project to study some of these problems.

This study addresses some of the problems of in the fisheries sector with a hope of making a small contribution to the sound fishery policy formulation. We hope that the knowledge generated through this study will help in some way or the other to redress the problems of the sustainable fisheries development.

Foreword

Fisheries sector is one of the important sources of earning income particularly for a large number of coastal people belonging to economically backward class. Fisheries sector plays an important role in the socio-economic development of the nation. It has a great potential to contribute to national income, food security and foreign exchange earnings. It not only creates direct employment opportunities in the sector but also provides employment in other allied industries.

Like elsewhere, fishing in India is one of the oldest occupations for the people living in coastal areas. However, in India fishing was not regarded as business. It was a way of life for a large segment of fishermen. The fishing was done at a subsistence level. The fishermen had generated wealth of knowledge about fish and sea through their experience of fishing for generations. The fishing was a traditional and family occupation. This kind of fishing did not pose any threat to the fishing sector.

However, in recent years, fishing activity all over the world has experienced a dramatic transformation in terms of mechanization, motorization, commercialization etc. In India too, fishing sector is subjected to this global pattern of modernization. The founding of World Trade Organization in 1995 opened up new vistas of growth in fisheries sector. This provided an impetus for an increase in production and commercialization of fisheries. Today, India is one of the top ten fish producing nations of the world. The introduction of new technologies, changes in the institutional arrangements and in the management of fisheries have made fish a tradable commodity at the national and international level.

These changes in fishing have opened up new challenges and issues. Indiscriminate exploitation of marine resources, depletion of precious marine resources, unemployment, replacement of old patterns of relations with individualistic and competition patterns, decreasing or stagnating marine fisheries production etc are some of the challenges that need immediate attention. The issue of sustainable development of fisheries remains a concern even today.

The present book entitled "Fishery Management" largely deals with the above mentioned issues. The issue of sustainable development of the marine resources calls for better and efficient management of the fisheries sector. The marine resources can become unsustainable if they are not properly planned and managed. The State is on the forefront in all the efforts to achieve sustainable development of fisheries. The State as a manager of the resource can ensure sustainable development through appropriate management techniques. However, the other stakeholders in the fisheries like fishermen have equally important role to play in sustainable marine fish production.

The Chapters in this volume look at fishery sector broadly, covering the historical aspects of fishing, mechanization of fishing and the conflict between the traditional and mechanized fishermen. The book has also focused on Management of Fisheries, which is the main concern today. The book organized in Chapters covers topics like role of women in fishery sector, economics and marketing of fisheries, impact of mechanization on fishing, conflict, management of fisheries, etc. The authors without indulging in jargonistic writing have tried to explain the matter in a simple and natural manner. All the Chapters in the book are worth studying.

This book based on the study of fishery sector in Goa is first of its kind. No such academic endeavour has been done to understand the fisheries. The book is based on inter disciplinary approach covering the social, economic and political aspects. It is an important contribution to the understanding of varied aspects of fishery sector in Goa. I am sure, this book will be of great use to the researchers, policy makers and to all those engaged in the study of fishery sector.

I congratulate the authors of this book for their keen efforts in studying, compiling and analyzing the various issues related to the fisheries in the State. I must also congratulate the MES College for its role and keenness in bringing out this book.

The inspiration and initiatives generated will be permanently enshrined in the development of the fisheries sector in the State.

Shri. S. C. Verenkar
Director,
Directorate of Fisheries,
Government of Goa.

ACKNOWLEDGEMENTS

This major research project is the out come of unstinted co-operation and help rendered by many. We extend very special and heartfelt thanks to all of them.

We are highly grateful to the University Grants Commission, New Delhi for accepting our research proposal and sanctioning the research grant to execute the project. The UGC authorities/officials were kind enough to answer our queries raised time and again regarding grants and other matters.

We owe a special debt of gratitude to Dr. Subramanian, principal scientist of fisheries, Indian Council of Agricultural Research (ICAR), Ela, Old Goa, Dr. Ramachandra Bhatta, professor, department of fisheries economics, fisheries college Mangalore, Dr.J.K.Stephen, HOD, Economics, S.T. Hindu College, Nagarcoil, Dr. John Kurien, Director, Centre for Development Studies, Cochin, Dr. M. E. John, Fisheries Survey of India, Vasco for their valuable guidance, suggestions and help in many ways.

We also express our gratitude to the Libraries of several universities and institutions for the collection of secondary data. Some of them are; CMFRI, Cochin and Mumbai, Fisheries College Mangalore, Mumbai University, Mumbai, Goa University, Goa, ICAR, Ela, Old Goa etc. We are grateful to the librarians and other staff of these libraries for helping us in getting the necessary and relevant information. We are thankful to ex-Librarian of our college Shri Vijay Hazare and other staff members of library for providing us all the available information on fisheries in the library.

We are extremely thankful to Shri. S. C. Verenkar, Director of Fisheries, Government of Goa for writing Foreword for this book. The Director and other officials from the Directorate of Fisheries and fish surveyors in selected villages were kind enough in providing us the much needed primary data pertaining to fisheries sector in Goa.

We sincerely thank all fisher folk respondents with whose co-operation this research project was completed. They were kind enough to spend their valuable time to respond to the lengthy and time consuming interview schedules.

We express our deep sense of gratitude and thanks to Miss Shashikala Sarmalkar and Miss Kaveri Kumbhar, project fellows, for assisting us in all possible ways even some times going out of their way and working systematically and tirelessly for the completion of the project report.

We are much obliged to our founder principal and vice president of M.E.S. society Shri M. S. Kamat, ex-principal Shri D.A. Kamat, present principal Dr. R. V. Hajirnis for their valuable co-operation during different stages of this work. We also thank Mrs. Meenakshi Bawa, Mrs. Harsha Shirwaikar, Shri B.S. Ingalhalli, Shri Amey Kuncoliekar, Shri Bhagvan Morje and Shri Vishnu Kerkar for their timely help. We thank our office staff for their co-operation. Finally, we thank each and everyone who helped us directly or indirectly in the completion of this project report.

Contents

List of Tables

Map Of Goa Showing The Selected Fishing Areas

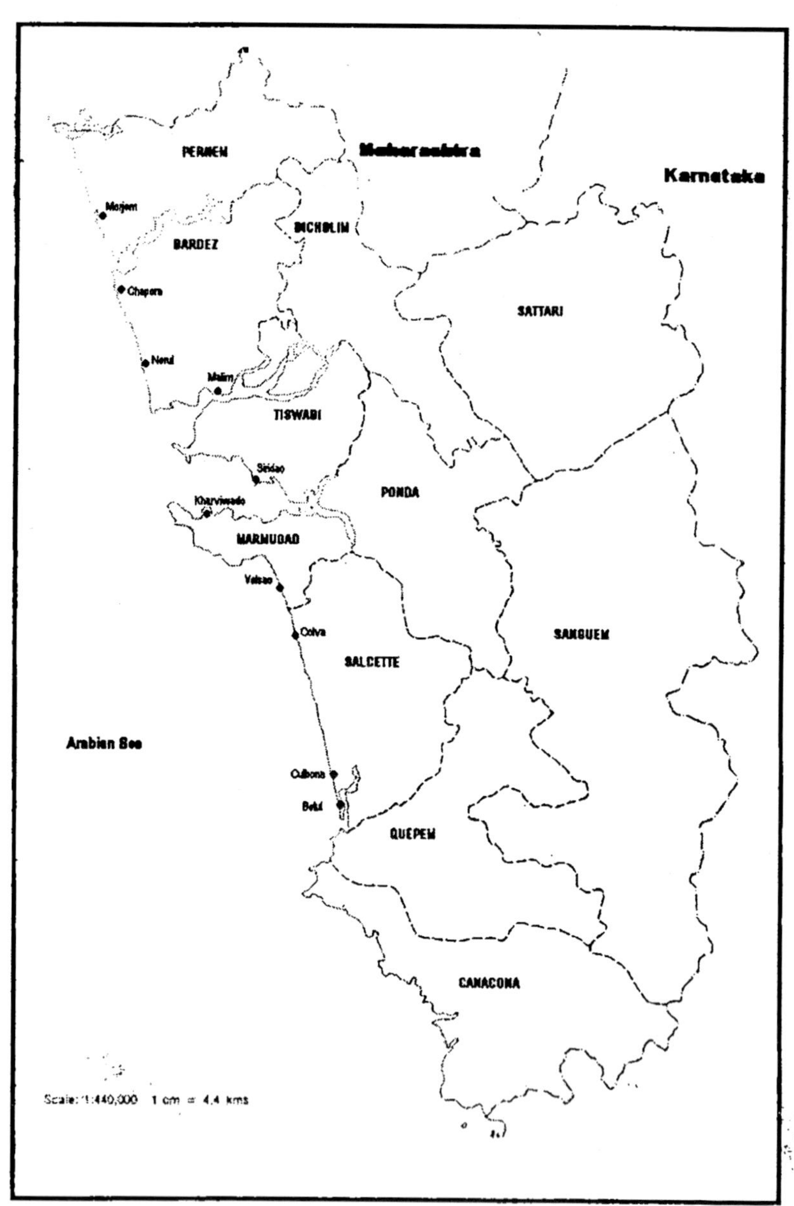

CHAPTER I

INTRODUCTION

BACKGROUND OF THE STUDY

Fish has been the most important and sustaining factor for the people around the world. It is particularly true in case of the coastal communities, as fish constitutes the most important part of their diet and also the main source of their livelihood. Fish is a major industry for the coastal people. It is a main raw material for many industries – the manufacture of fishmeal for the poultry and the production of oils of various kinds.

However, in recent years, the commercialization and mechanization of fishing and resultant intensive fishing and ruthless exploitation has put tremendous pressure on the inexhaustible and otherwise renewable resource. There is a cliché, "too many fishermen chasing too few fish". So the important issues of concern are how to ensure the optimum and the best use of the existing stock of fish and an uninterrupted and sufficient supply for the future generation. These two questions get the centre stage on the agenda of the ecologists and of the fishing communities all over the world.

The major issue is not only to have a proper supply of fish but also to share the available catch among the different groups of fishermen. This necessitates an international as well as a national regulation of the method of fishing. The problem is of such a great concern that international agreements have been reached on the fishing regimes on the use of nets of large mesh so that mature fish can be harvested while it's juvenile is allowed to grow. Efforts are also made to protect spawning stock and fish eggs.

The need for a sound fishery management policy is precipitated by the following three main considerations:

(1) Over exploitation or depletion of the stock to biologically harmful levels will result in a loss of

potential benefits such as food, income and employment, both in the long-term and short-term;

(2) Ecological damage that may result in negative effects on the fish population itself and also on other species in the habitat. This is because a very low level of any stock is likely to have negative impacts on other dependent stocks; and

(3) Economic waste due to over investment in the fishing activity causing over capacity and resulting in the loss of future income (Sandy: 2001).

The State of Goa is no exception to this kind of predicament. Fish is an inseparable part of life of the people of Goa. Goa has a long coast line of 105 kms and around 250 kms of inland waterways and 4000 hectares of marshy lands along the estuaries. With the commissioning of Salaulim and Anjunem irrigation projects and around 100 hectares of fresh water tanks, the total fresh water resources of Goa are around 3800 hectares and around 30225 people are dependant on fishing for their livelihood (Directorate of Fisheries: 2004).

This clearly brings out the importance of fish in Goan social and economic life and also in their diet. Pulses and vegetables form very little portion of the Goan diet. Milk is also not a popular consumption. Hence, fish is the only diet providing the required quantities of protein to the people. Table 1.1 shows the protein content of fish commonly consumed in Goa.

Table 1.1 - Protein content of fish commonly consumed in Goa.

Name	Protein (per 100 gms of edible portion)
Mackerel (Bangada)	18.9
Sardine (Pedve) low fat	21.0
Sardine (Tarle) high fat	20.5
Small prawn (Sungata)	19.1
Crabs	11.2

Source: Nutritive value of Indian Food; NIN, Hyderabad, quoted in a Citizens Report on the State of the Goan Environment (1993) entitled, Fish, Curry and Rice, Mapusa: the Other Indian Press.

As fish is an important part of Goan diet naturally fishing, is a major occupation of a large section of the Goans, leading to many other allied activities. The preservation of fish is a traditional household industry. Besides, people are also engaged in net repair. The members of the fishing community are engaged in making of fish pickles, balchao and parra.

Since liberation, Goa's fishery has undergone rapid growth and development. From the traditional fishing, Goa has gone to mechanized fishing – trawling and purseining. Fishing in Goa has been commercialized. However, mechanization of fishing did not reach majority of the traditional fishermen known in the local language as 'ramponkars' and 'pagelkars'. Though, a few of the traditional fishermen bought mechanized boats, as per the data obtained from the office of the Directorate of Fisheries nearly 1035 mechanized vessels of 30–120 H.P and above operate in Goan waters. They are either using trawler nets or purse seines.

The entry of the mechanized boats into this area made the traditional fishermen raise the battle cry that their livelihood was threatened as the trawlers were operating in the same zone in which the traditional fishermen operated. To protect their interest they formed an association – Goencho Ramponkarancho Ekvott (GRE). A bitter conflict emerged between these two groups of fishing communities. Both looked to the Government for arbitration.

A Citizens' Report on the Goan Environment (1993) pin points the impact of industrialization, commercialization and mechanization on fishing as "Goa, after liberation inevitably got caught up in the process of development which was taking all over the country. Development had a major impact on the whole fishing network in Goa." Increasing industrialization resulted in considerable marine pollution. There was a massive increase in the number of mechanized fishing vessels resulting in displacement of traditional fisher folk. A major conflict over resource emerged between the traditional fishermen and the owners of mechanized fishing vessels. A people's struggle grew around the problems of traditional fishermen.

The main focus of this book is on the socio-economic and political impact of development process on fish harvesting and on fishing community. It attempts to emphasize both the changes and continuity aspects of the fishing community and the communities involved in it. It tries to explore the unanticipated consequences of mechanization of fishing and how the traditional fishermen have adapted to the process of commercialization and mechanization. Mechanization has generated a lot of discontentment among the traditional fisher folk. This has given rise to conflict between traditional and modern fishermen. The book seeks to explore and analyze the process of transformation of this discontentment into a movement.

As the authors belong to different disciplines an attempt is made to analyze the socio-economic and political aspects of the problem, the study also looks into the nature of social structure of fishing community, the recent movement of the traditional fishermen, the impact of political decisions on the fishing community and economy and sustainable development of fishing industry, etc., are some of the issues covered in this book.

REVIEW OF LITERATURE

A review of previous studies on fisheries is essential to get a bird's eye view about the sector. The information gathered is useful to go in depth and to identify the unknown areas. Several authors have dealt extensively with various aspects of fishery sector. Some of the studies are reviewed in this section.

John Kurien (1982) has studied "Technological Change in Fishing and its impact on Fishermen in Kerala. He found that there was a considerable increase in production of fish in Kerala. The value of output increased phenomenally and 90 per cent of the increase was attributable to price change alone. He argues that the mechanised sector received the major share of the benefit of the increased output. Private entrepreneurs received the bulk of the share while the working fishermen reaped very little benefit.

Subba Rao's (1986) study "Economics of Fisheries" reveals that the quantum of financial resources allotted to the fisheries sector in general and to specific schemes in particular is very meagre. Although, a number of schemes have been proposed in each five year plan majority of them were not completed in the respective plans. There is a lack of coordination among various agencies connected with the implementation of the different fisheries programmes to raise the level of fish production.

U.K. Srivastava and M. Dharma Reddy (1986) in their edited work entitled "Fisheries Development in India" opine that fishery is a potential employment generating sectór and a source of animal protein at low prices. The book focuses on the inter-relationships between production, marketing and organization. The authors describe the different aspects of fisheries development like demand, supply, marketing infrastructure, investment opportunities, technology and organization.

Gracy's (1988) study entitled, "Impact of technological advancement on socio-economic conditions of women in fisheries in Kerala" attempts to highlight the role of women in fisheries and the impact of technological advancement on socio-economic conditions of women. The technological advancement in fisheries has brought a lot of change in the socio-economic conditions of fisherwomen adversely affecting their livelihood. The modernization also resulted in replacement of indigenous handmade nets by machine made nets. The study reveals that no attention has been given to acknowledge and understand the role of women in fisheries.

Perviz De Souza (1988) in his study "Caught in a net" conducted in Goa observed that mechanization of fisheries in 1960's has increased fish catch in Goa. In the process of increasing fish catch the Government has ignored the marine ecology and traditional fishermen.

Mechanization of crafts was achieved by providing subsidies and loans but ancillary facilities like berthing for various crafts, training programme, ice plants and cold storage wings took a long time to develop and are even today insufficient. Similarly, the plants and technology for extraction of fish oils and fish meal have been virtually neglected. He also claims that the data relating to fishery has not been properly maintained.

Rao P.S. *et al.* (1988) studied "Wages and wage structure of fishing operations: "A case study of Versova fishing village". They made an attempt to study the working conditions of fishing workers employed in different types of fishing boats and the pattern of wage structure. The survey was undertaken from September, 1984 to June, 1985. Their study revealed that in this village wages were generally high due to the absence of system of collective bargaining of wages. If the workers organize themselves into unions the uniform wage structure would help the fishing industry to stabilize and promote their interest.

Panikkar K. K. P. and Sathiadas R. (1989) in their paper "Marine fish marketing trend in Kerala" deal with the fish marketing system prevailing in Kerala. Their study indicated that due to lack of infrastructure facilities the supply of fish at the landing centre is highly inelastic which often would be result in disposal of fish at throw-away prices at the time of heavy landings. The involvement of a number of middlemen in the marketing chain adversely affects the interest of both the fisherman and the consumer. This can be rectified only through government interference by announcing a support price for those varieties which used to be caught in large quantity.

J. Vasanthakumar and K. R. Sundaravaradarajan (1990) in their paper entitled "Adoption of Scientific Technology by Trawler operators of Tamil Nadu" discuss about various aspects relating to adoption of scientific technology by trawler operators of Tamil Nadu. The study was conducted in six maritime districts of Tamil Nadu. The study found that the trawler operators face the problems such as inadequate finance, lack of berthing facilities, non-availability of fuel, etc. But non-availability and underweight may be tackled through effective intervention by co-operative societies. The authors conclude that the State Department of Fisheries and co-operative societies may evolve an extension strategy with emphasis on radio and personal contacts for effective diffusion of scientific technology together with short-term training programmes.

Sathiadhas R. and Panikkar K. K. P. (1992) in their paper "Share of fishermen and middlemen in consumer price: A study at Madras region" discuss the marketing margins, and producers' and middlemen's share in consumer's rupee for commercially important varieties of marine fish in Madras region of Tamil Nadu. The data has been collected for 15 to 20 days in each quarter during 1984-85. On the basis of the survey they conclude that to protect the interests of both producers and consumers it is necessary to reduce the magnitude of marketing margins. By introducing co-operative marketing system, involvement of too many intermediaries can be avoided to increase the efficiency of fish marketing system.

Ibrahim P. and D'silva S. (1994) have studied "Economics of Mechanized Boats and Motorized Crafts". They used efficiency measures like catch per trip and effort, labour productivity, capital intensity and cost effectiveness in different crafts-gear combinations like odam-boat seine, thoni, gillnet and trawlers. The thoni-gillnet combination yields a higher net profit than others. Their conclusion is that mechanized boats (trawlers) neither affect a bigger output nor make a larger profit than the motorized crafts. This certainly calls for a shift in the mechanization policy of the government.

Sathiadas R. and Kumar Narayana (1994) studied "Price Policy and Fish Marketing System in India". The authors strongly feel that the growth of fish production and development of fishery sector is highly dependent on an efficient fish marketing system. The pricing efficiency is concerned with improving the operation of buying, selling and connected aspects of marketing process so that it will remain responsive to consumer behaviour. They emphasized that modern fish marketing policy should envisage not only meeting the existing demand for fish but also tapping the potential demand.

Singh, Katar (1994) studied "Marine fishermen co-operatives in Kerala". The study revealed that the socio-economic condition of most of the fishermen in Kerala has not improved significantly as a result of the establishment of fishermen co-operatives. The co-operatives have helped their member fishermen in the motorization of their boats and in acquiring modern nets and other supplies.

But most of the member fishermen do not believe that the co-operatives are their own organizations established to promote their socio-economic well-being. There is no change in the borrowing pattern of the fisher folk after the establishment of co-operatives and the entry of banks in the sector. The co-operatives can win their co-operation and loyalty if they intervene in a big way in fish processing and marketing and assure the fishermen a higher share in the consumer's price.

The study by Silas E. G. (1995) on "Conservation and Regulation in Marine Fisheries" indicates that fisheries management should aim at a sustainable long-term economic utilization of the resources. This can be achieved by managing the fish stocks through proper regulatory measures and controlling fishery dependent factors such as access and efforts. He suggests that while ensuring biological productivity, the socio-economic and environmental and conservation issues also need to be addressed.

Kalia S.K. (1996) in his paper entitled "Financing Fisheries Sector" discusses about the fisheries scenario in the country and the credit aspects for specific projects. The NABARD has been providing grants from its Research and Development funds for conducting further operational research, standardization and commercialization of technologies. In consultation with ICAR it has also prioritized certain research areas such as cold water fish culture, development of suitable feed for high density culture of air breathing fishes, pen culture and cage culture of inland fishes, breeding of aquarium fishes, post harvest technology linked with deep-sea operations, utilization of bye-catch of trawlers, seaweed culture technique for prawn feed and standardization of culture, development of prawns other than tiger and white prawns. The author hopes that such funds would be available in future.

The study by Bhaumik U. and Saha S. K. (1998) on "Role of extension in arousing mass awareness and public participation in fish conservation movement" throws light on the role of extension, mass awareness and public participation in fish conservation movement. There has been severe decline in the fish fauna of inland waters of India due to anthropogenic interference, which calls for fish germplasm conservation. The strategy for transfer of technologies on fish conservation to the members of the target group is to be treated henceforth as one of the essential inputs in the overall activity of germplasm conservation programme. The prime objective of fish conservation programmes could be achieved only by giving importance to communication. The village motivators play key role in establishing good participation in the activities of fish germplasm conservation. Thus, fish conservation movement involves both individual and group action.

K.K.P. Pannikar, *et.al* (1998) have studied "Structural Changes in the traditional Fishery of Kerala and its socio-economic implications". They highlighted the socio-economic implications of the structural changes on the traditional sector. They pointed out that before motorization phase rural landing centres were primary markets for the traditional sector. Increased ring seine operation with its huge landings attracted many traders. As a result bargaining capacity of the traditional sector accelerated. The share of fishermen in consumer rupee has been increased. The study also pointed out that the fishing gears used by the fishermen are destructive posing the problem of conservation of fish resource. Better economic performance has resulted in increased size of craft and net as well as HP of engines gradually led to higher investment and operating cost.

Nair, Mini (1998) studied "Women in fisheries – Emancipation through co-operatives". The study revealed that fisherwomen are relegated by fishermen since they feel fisherwomen are involved in less productive areas like fish vending and other post harvest activities. Fisherwomen engaged in fishery related activities are marginalized due to technological changes. This has affected not only their individual income but also the income of their families.

Sutton Michel (1998) in his paper "Harnessing market forces and consumer power in favour of sustainable fisheries" suggests that greater public involvement in the fishery management process must be encouraged. Marketed economic incentives must be created to promote sustainable fishing. Conservationists working with responsible, progressive sea food companies and other stakeholders must develop reforms that will encourage fish buyers to purchase their products only from sustainable, well managed fisheries.

Verghese C. P (1998) has done a study of "Capture Fisheries Technology in India". The study revealed that the fishing industry is not able to harvest the resources because of the low value of resources in the deep waters and expensive operation of large vessels in this area. If the exploitation of the harvestable resources in the off-shore and deep waters has to be done, generation of fuel efficient vessels below 23m class with cost effective gears and methods is needed.

Sehara D. B. S. et.al (2000) studied "Economic Evaluation of Different Types of Fishing Methods along Indian Coast". They indicated that the studies on the craft and gear combinations should be conducted on macro level in different maritime states for effective future planning in maritime fisheries sector.

Non-mechanized, motorized and mechanized sector should be given equal priorities while undertaking cost and earning studies. There has to be also a stress on employment potential, marketing problems, financial needs of various types of units and proper management of inputs used for fishing.

Immanuel Sheela and Srinath Krishna (2000) studied "Potential Techno- Economic Role of Women in Fisheries". The study revealed that women contribute a lot to fisheries sector. In coastal areas, women play an important role in fisheries and in some parts of the world they are good navigators too. Modernization has diminished the role of fisherwomen but yet they play an important role in the fishing activity. The authors suggest that women should be helped to participate in production activities without disturbing their domestic responsibilities.

Kumar Suresh P. (2001) in "New Technology and Artisanal Fishermen in Kerala" makes an attempt to discuss the emergence of motorization and the marginalization of the traditional fishermen. The study reveals that modernization of fishing industry has led to marginalization of artisan fishermen.

Rajasenan D. (2001) in his paper "Technology and labour process in marine fishery: The Kerala experience", believes that the amount of fish that can be harvested per unit of time depends primarily on the level of technology, which in turn depends upon the size of investment and the number of fishermen that participate in fishing operations. Mechanization has resulted in higher production and productivity. The new labour process in mechanized fishing due to capital intrusion has brought a change in the theory of labour process. It also resulted in outboard motorization.

Sathiadas R. and Kumar Narayana R (2001) in their paper entitled "Export of finfish- impact on domestic trade and production" assess the trends in production and export trade of selected varieties of finfish based on the data collected from the publication of CMFRI and MPEDA, Cochin. Their study reveals that in spite of exporting the top quality selective fishes, the unit value realized by them from abroad is not appreciable and even less than that of the prevalent domestic prices for some varieties. Expansion of export trade of finfish without enhancing the internal supply of quality fishes will be detrimental to the interest of domestic consumers. Aquaculture alone is the viable alternative for the same. Diversified fresh and brackish water aquaculture production of quality fishes and sea farming should be intensified to bridge the gap between demands and supply in the domestic market and to maintain the tempo of export trade of fin fishes.

Vijayakumaran K. (2001) in his paper "Social Audit - an ideal method for neutralizing conflict situations in aquaculture industry" attempts to identify and examine the activities of the aquaculture industry having an impact on the immediate environment and to elaborate how social audit would help to neutralize the conflict situation in aquaculture industry. The author states that once the factors contributing to the social costs/benefits are identified, their objectives, assessment, evaluation, measurement and presentation will give an idea of the net cost/benefit of the farm. The study reveals that the understanding of the principles and adoption of the practices will certainly lead to the achievement of sustainability in the long-run.

Barbosa (2002) in his study "Fishing for a High Living" states that the mechanised fishing in Goa is done without any proper regulations. There is no license system for trawlers in the State. Once a trawler owner registers his trawler, he need not approach any government department again. The state of Goa has 1128 registered trawlers and this is far above the saturation point. He suggests that there is a need to regulate the number of trawlers that go in the sea and their expedition schedule. The author strongly feels that there is a need to redraft the laws and to control the mesh size.

D'Souza (2002) in his study "Fishing Woes to the Fore" asserted that the main reason for the malaise in the fishing scenario in Goa is the lackadaisical approach of the governmental agencies, both Central and State. The Central Government has still to enact the law to implement a uniform ban period on fishing activities all over the country. He stressed that there should not be more than 800 trawlers operating if there is to be optimum fishing yield in the state. He found that pollution by organic releases generated from industrial units, urban settlements, hotels and shrimp farming activities have adversely affected bio-productivity in Goa's coastal waters. His main conclusion is that the quantity and quality of fish is slowly deteriorating in the state.

Tiwari P. C. (2005) in his paper "Natural Resources Information System for Wasteland Development in Uttaranchal: A Planning Support System for Sustainable Watershed Management in Uttaranchal Himalaya" concludes that the sustainable development of wasteland development strategy is community and people oriented, and the local people are willing to adopt the practice of sustainable resource development and utilization.

Venugopal S. (2005) in his book "Aquaculture" claims that information and consumer education programs can play a vital role in expanding the demand for aqua cultural products. He suggests that economists should be innovative in their research approaches to study consumer demand for fish and sea food and also about the resource allocation and public policy affecting aquaculture.

It has been noted that most of the studies are related to other parts of India. Till date there has not been any comprehensive and exhaustive study on Goa's fishery sector. It is hoped that this work would be an humble attempt in this direction.

Objectives of the study

The main objectives of this study are:

1. To trace the growth of fishery sector in Goa.
2. To study the socio-economic conditions of the fishing community in Goa.
3. To examine the role of women in fishing sector.
4. To understand the economics and marketing aspects of fisheries.
5. To examine the impact of mechanization on the fishing community.
6. To study the conflict between the mechanized boat owners and traditional fishermen.
7. To examine the Fishing Regulatory Regimes adopted by the Government – its implementation and shortcomings.

8. To study the surveillance measures taken by the Government of Goa to monitor the "poaching" and over exploitation of fish resource.

9. To analyze the Conservation and Sustainability aspects of fish resource and to suggest a sound management policy.

METHODOLOGY

The Goa State has a long coastline of about 105 kms. The fishing activity is carried out all-along this coastline. Therefore, the sample areas selected for study represent the entire State of Goa. The State of Goa is divided into two Districts namely South Goa and North Goa. The list of fish landing centres in each of these districts is shown in table 1.2.

Table 1.2 Fish Landing Centres

Districts	Fish Landing Centres
North Goa	Caranzalem, Dona Paula, Siridao, Arambol, Anjuna, Sinquerim, Vagator, Assago, Siolim and Verem.
South Goa	Vasco, Cansaulim, Velsao, Majorda, Colva, Mobor, Betul, Palolem, Khola, and Agonda.

Source: Directorate of Fisheries, Panaji, Goa

From the above-mentioned centres, five centres from each district were selected for the study on the basis of purposive sampling method. Some important criterion used in selecting the fish landing centres are: maximum fish catch as per the statistics available from Directorate of Fisheries collected at the time of the commencement of this study. Moreover, they fairly represent the entire Goa. The names of the fish landing centres selected for the study are shown in table 1.3.

Table 1.3 Fish Landing Centres

Districts	Name of the areas selected for study
North Goa	Chapora, Malim Jetty, Morjim, Nerul, Siridao
South Goa	Kharviwado, Velsao, Betul, Cutbona, Colva

BRIEF DESCRIPTION OF SELECTED FISH LANDING CENTRES

Chapora

Chapora, which lies on the north coast of Goa is 10 km from Goa's commercial town of Mapusa. Dependent on fishing and boat building, it has, to a great extent, retained a life of its own independent of tourism.

There are around 1000 households in Chapora. The total fishermen population is 5,500 but only 4100 are actually involved in fishing. There are altogether 65 mechanised trawlers and 15 registered canoes in Chapora.

Malim Jetty

Malim Jetty is seen across the river Mandovi. There are around 2000 households at Malim. 270 trawlers are operating from this jetty though many of the trawlers stay away from Malim. Total fishing population is around 7000 at Malim of which active fishermen are 4500.

Morjim

Morjim comprises a pictures portion of Goa's 103 square metre long shoreline in the North. The water of the Arabian Sea and the Chapora River provides a rich breeding zones for a variety of fish. The people are involved in artisan fishing.

There are 25 registered canoes at Morji. Some fishermen from Morji also own trawlers, but they operate from Malim Jetty. Morjim is a fishermens village. As the legend goes goddess Bhagwati was fished out by the *morjes* (fishermen) came to be called Morzai. The goddess is believed to have come riding on a *mhor* (peacock) and hence *mhor-jim*. The peacocks which frequent the hillocks amidst the cashew plantations are revered creatures.

Back ashore, they recount legend after legend while they mend the fishing nets. One such legend claims that the village was being gobbled up by the sea. But a *satpurush* (a holy person) implored mother nature and she saved the land. The grateful villagers erected the Satpurush temple at the foothills, venue of the *divzam* - a festival in which people dance with burning torches on their heads.

Nerul

Nerul located in North Goa is a pictures village. It shares its boundaries with the historic village of Verem and Reis Magos. The green hills, the blue water of river Mandovi and the cool breeze of the Arabian Sea made Nerul a hot favourite of the Portuguese officials during the colonial days. The portuguese gently come to the Nerul beach to escape from the heat of the Goan Summers. With the launching of modern tourism the Nerul beach has become a favourite picnic spot and salt water bath both for the locals and tourists.

The population of Nerul is predominant by Catholic. Though, agriculture is the basic activity of the village, fishing is an important occupation of the people. There are 150 registered canoes at Nerul. The total fishing population is around 700 and the number of active fishermen is 200.

Siridao

Siridao is situated near the Zuari estuary. Siridao Beach is a shell collector's haven with its assortment of oyster and pearl shells. There are around 25 canoes - traditional crafts in this village. The total fishing population in this village is 400. Some fishermen from this area own trawlers. But their trawlers are kept at Malim Jetty as there is no jetty at Siridao.

Kharvivado

Kharvivado falls in Vasco-da-Gama which is a key shipping center. Vasco da Gama has quite an extensive fishing fleet and boat building/shipyard located in and around the harbour. Kharvivado which is known in local language as kharevaddo is a fishing area. There are 200 mechanised trawlers and 40 motorised canoes. Total fishing population of Kharvivado is around 200.

Velsao

Velsao is a village in Mormugao Taluka. There are 25 registered canoes in Velsao. Some fishermen have fitted motors to their canoes. Total fishing population of Velsao is around 500.

Betul

Betul is a place located in the Salcete side of Goa. This place is a one hour distance from Margao the commercial capital of Goa. Betul is known for its beautiful Betul beach is situated along the southern part of Goa. Betul beach comparatively a small beach, it is located along the southern end of the Sal River in Goa. Fishing is the primary occupation here. The fishing population is around 400.

Cutbona

Cutbona is one of the important jetties in South Goa district. There are 190 mechanised trawlers operating through this jetty. The fishing population at Cutbona is around 250.

Colva

Colva Village lies nearly 6 km away from Margao. The 25 km of coastline makes Colva important fishing village. There are 40 mechanized trawlers and 25 canoes in Colva. The fishing population is around 700.

The Colva beach starts from Bogmalo in the north to Cabo de Rama in the south. The Colva Beach was once the favourite weekend getaway for the upper class elites of Margao. They often headed to Colva for the "Mundanca" or the 'Change of Air'.

SELECTION OF THE RESPONDENTS

This study is based on the information elicited from 400 respondents. These respondents belong to the traditional, motorized and mechanized fishermen categories. There is no reliable and updated list of these fishermen. Neither the government nor the fishermen's associations have maintained such a list. Moreover, the fishermen are not easily available due to their long hours of fishing activity and rest thereafter. Therefore, we had to rely upon the purposive sampling method for selection of the respondents. However, enough care was taken to have the heterogeneous representation of the fishermen in terms of age, education, religion, caste, craft, etc.

One finds substantial variation in the number of traditional, motorized and mechanized fishermen in the areas chosen for the study. They are not evenly spread in all the areas. Secondly, in recent years there has been decrease in the number of traditional fishermen as they have shifted to motorized fishing. Hence, one finds variation in the number of fishermen (respondents) chosen from the traditional, motorized and mechanized categories for this study.

TOOLS OF DATA COLLECTION

Primary Data

Interview with the help of structured schedules, as a tool of data gathering was used extensively to collect relevant information. Besides, observation technique was used as a supplementary method to collect the required data. Questionnaire method was also used to collect information from Ministers, concerned Government Officials and activists relating to the fishing industry.

Secondary Data

The necessary secondary data has been collected mainly from the official records both published and unpublished from the Directorates of Fisheries and Planning and Statistics. Besides, books, newspapers, journals, magazines and websites on internet were referred to collect the requisite information.

SIGNIFICANCE OF THE STUDY

This study will be of interest to academicians as well as to the general public because it is a first study of its kind. It endeavours to make a comprehensive study of the fishing activity in Goa from socio-economic and political perspectives. The topic under study is contemporary and needs immediate attention of the academicians to find solutions to the emerging problems. The simmering discontentment among the traditional fishermen on account of varied problems faced by them, may take violent turn if proper remedies are not taken. This study can contribute to the holistic understanding of the problem and to formulate a programme to redress the grievances of the fishing community in Goa. The study can make a significant contribution to identify the genesis of the problem.

This study will generate information which will be a milestone in understanding one of the dominant communities of Goa. The other significant contributions are as follows.

(1) This study is aimed at drawing the attention of the Government to the conflict between Ramponkars and Mechanized fishermen. It would impress upon the Government the need to come out with a concrete policy of arbitration between the two rival groups, and review the fishing policy.

(2) The findings of this study will help to evolve some concrete fisheries policies. Such fisheries policies might give broad direction and priorities on how the resources of a nation or region are to be utilized. This would go a long way in achieving the goal of sustainable fisheries.

(3) This study would expose the plight of the traditional fishermen and might motivate the Government to provide some relief to the traditional fishermen.

(4) The present study might be of a great interest to the environmentalists and public at large in knowing about the monitoring, control and surveillance systems for the capture of fisheries.

LIMITATIONS OF THE STUDY

This study is of descriptive nature and is based on both the primary and secondary data. This study has been handicapped by the fact that the relevant data was not available with the concerned departments. These offices have expressed their inability to provide the necessary information due to inadequate manpower. As this study is also based on the primary data we cannot claim 100 per cent accuracy. In spite of our ardent efforts to get accurate information, it was seen that many of the traditional fishermen were ill informed and those who were knowledgeable were unwilling to give the information, which they felt could be detrimental to their interests. Many others are not in the habit of keeping systematic records either about their income and expenditure or about the cost and benefits. It is these difficulties that have hampered the scope of this study.

CHAPTER II

Fishery Sector in Goa—Facts and Figures

Goa the twenty-fifth State of the Indian Union is situated on the west coast of India geographically situated between 15° 48'N 14° 53' N latitude and 74° 20' E and 73 ° 40' longitude (Angle: 2001). The altitude of the towns of this state ranges between 20 and 62 meters above the sea level. It has a coastline of 105 kms. Goa has a warm tropical climate. It receives heavy rains from June to August. The average rainfall is 305 cm with the temperature ranging from 25-35 degree Celsius.

As per 2001 census, the population of this state is 13,47,668. The population of North Goa district was 7,58,573 and that of South Goa district was 5,89,095. As per the census, the density of population is 364 persons per sq. km, which is higher than that of the all India level, and the neighbouring states of Karnataka and Maharashtra.

The territory of Goa is cut across mountains, lakes, streams and beaches. Its main rivers are Mandovi, Zuari, Chapora and Sal. Mormugao harbour in Goa is one of the best natural harbours on the west coasts of India. It handles large exports of iron ore. The lush green meadows and the huge coconut palms gently swaying on the beaches, form a part of the Goan landscape.

Goa's scenic beauty and the distinct cultural heritage and stable climatic conditions have made Goa a major tourist centre. Every year thousands of tourists both foreign and local visit Goa. Thus, tourism has become an important industry and occupation for hundreds of people. Almost 70 per cent of the population in the state depends on tourism-related sector.

As in the rest of the country, agriculture is still a very important occupation of the Goan people. The introduction of tractors, power tillers, and threshers has led to an increase in the agricultural yield. Rice is the staple food of the Goans and it is cultivated both in the kharif and rabi seasons. A number of cash crops like coconut, cashew, sugar, areca nut and groundnut are also grown. Cashew is an important crop and is a earner of sizeable foreign exchange. Cashew fruit is used in the production of local drink called "Feni".

Other major occupations of the Goans are farming, mining, dairy, animal husbandry, etc. Goa doesn't have heavy industries but in the post-liberation period Goa has witnessed the spread of many industries. The number of small scale, cottage and tiny industrial units has crossed 6000 mark. The number of cottage, medium and large scale units rose from a mere 46 units in 1961 to around 150.

The State of Goa is richly endowed with the industrial minerals like iron ore, manganese ore, bauxite, limestone, dolomite etc. However, the chief minerals exported by the state are iron ore, manganese ore and bauxite. Thus, mining is an important occupation of the Goans. In spite of industrial and tourism growth in Goa, the mining industry still continues to be the predominant factor of Goa's economy as it is the single largest contributor to the state GDP.

Its long coastline of more than 100 kms offers Goa an offshore fishing area of about 5200 sq. kms up to a depth of about 200 fathoms. With 250 kms of inland waterways, Goa is rich in fish resources. About 100 hectares of freshwater sources like lakes, Ponds and tanks can be exploited for the purpose of fisheries.

This has provided a vast potential for fishing activity in Goa. There are 8 talukas and 89 villages (42 marine and 47 inland) involved in fishing. In 2004, the fisheries population was 30,225 and the active fishermen population was 11,944 (Directorate of fisheries, Govt of Goa, 2004). Table 2.1 gives information about Goa fisheries.

Table- 2.1 - Goa Fisheries at a Glance 2004

1	(a) Coastal Length of Goa (kms)		104
	(b) Continental shelf (up to 100 fathoms depth)		10,000
2	Inland waterways in Goa (kms)		250
3	Inland water tanks (Ha)		100
4	No. of Fishing talukas		8
5	No. of Fishing Villages (marine)		42
6	No. of Fishing Villages (Inland)		47
7	Fisheries Population		30,225
8	Active Fishermen population		11,944
9	No. of Mechanized Boats up to 2004		1,152
10	No. of Motorized country crafts		1,035
11	No. of non-motorized country craft		1,079
12	No. of Registered nets		5,532
13	Fish landing centres (jetties)		5
14	Fish landing centres ramps		23
15	H.S. Diesel outlets operated through fisheries Cooperative Societies		4
16	Fisheries Cooperative Societies		6
17	No. of Members in Cooperative Societies		889
18	Annual Fish Landing (in M.T.) during 2004	Marine	84,394
		Inland	4,397
		Total	88,791
19	Value of fish landing (in M.T.) during 2004	Marine	17,196
		Inland	2,211
20	Export of Marine Fish Products, (in M.T.)		8,855
	Value (in lakh Rs.)		3,907

Source: Directorate of Fisheries, Panaji – Goa.

Table- 2.1 gives a bird's eye view of the fishery sector in Goa. With a coastal length of 104 kms and a continental shelf of 10,000 fathoms depth, Goa's fishermen population is around 30,225. Out of the 11 talukas in Goa, 8 talukas are actively involved in fishing. Goa has five fish landing centres known as jetties to facilitate the fishermen.

Since time immemorial fishing has been one of the major occupations of the people of Goa. In fact fish has become an integral part of life of the people. Fish is a staple food for about 95 per cent of the total population of Goa. It has provided quality food and sustainable income to thousands of people for centuries. Like agriculture, fishing in Goa was rather a way of life as it was done mainly for consumption purpose. It was not viewed as a business though it was a source of livelihood for many. The traditional fishermen were involved in subsistence and sustainable fishing. The traditional methods like Rampon and Gillnets were used in fishing. As a tradition, there was a total abstinence from fishing activity during the major part of the monsoon. Thus, the traditional fishing was in a perfect harmony with the eco-system.

The fishing not only provided a direct source of employment but it gave rise to many other allied activities. The members of the traditional fishing community were engaged in making fish pickles, balchao and parra. The preservation of fish was another traditional household industry. Many were also engaged in net repair (Ecoforum, 1993:82).

Barring few trawlers engaged by the Portuguese Government, fishing activity before liberation was carried out with non-mechanised boats (Angle, P.S. 2001: 87). However, with the liberation of Goa in 1961 the process of development began in all the sectors including fishing. Since then, the fishing sector has undergone rapid growth and development. During the last 2-3 decades the fishing sector has been subjected to intensive and rapid mechanization. With the emerging new opportunities for export trade and higher profitability on investment, more and more mechanized vessels were added year after year, taking advantage of institutional credit and liberal subsidies, concessions and other facilities by the Government. This is clear from the fact that the number of mechanized boats has increased from around just 4 in 1961-62 to

1128 in 2001-2002. On the contrary, the number of country crafts has come down from 4125 in 1961-62 to 1963 in 2001-2002 (D'Souza Joe, Goa Today, Nov. 2002: 17). In 2004, the number of country crafts was 1079 (Directorate of fisheries, Government of Goa).

The Table 2.2 gives information about fish production and mechanized and country crafts.

Table 2.2 Fishing Vessels and Fish Catch

Year	Fish production (in Tonnes)	Estimate Value (in Rs. Lakh)	Mechanised Crafts	Country Crafts
1961-62	17,000	78	4	4125
1970-71	36,616	900	110	3600
1980-81	25,715	985	186	2887
1990-91	56,225	2850	707	2000
1992-93	97,014	4750	764	2000
1995-96	85,418	5400	880	1980
1997-98	94,547	6608	1056	1840
1998-99	70,710	10,061	1092	1890
1999-2000	63,440	9068	1092	2194
2000-2001	69,386	15,793	1128	1963
2001-2002	73,135	16,544	1128	1963
2002-2003	83,756	19,407	1134	1963
2003-2004	84,394	19,786	1134	1963

Source: 1. D'Souza, Joe. 2002. 'Fishing Woes to the Fore', Goa Today, Vol. XXXVII, No. 4, Pg. 17.

2. Directorate of Fisheries, Government of Goa.

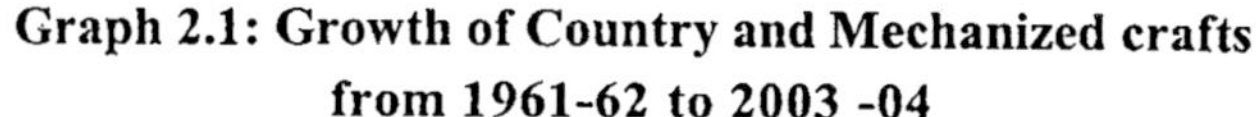

Graph 2.1: Growth of Country and Mechanized crafts from 1961-62 to 2003 -04

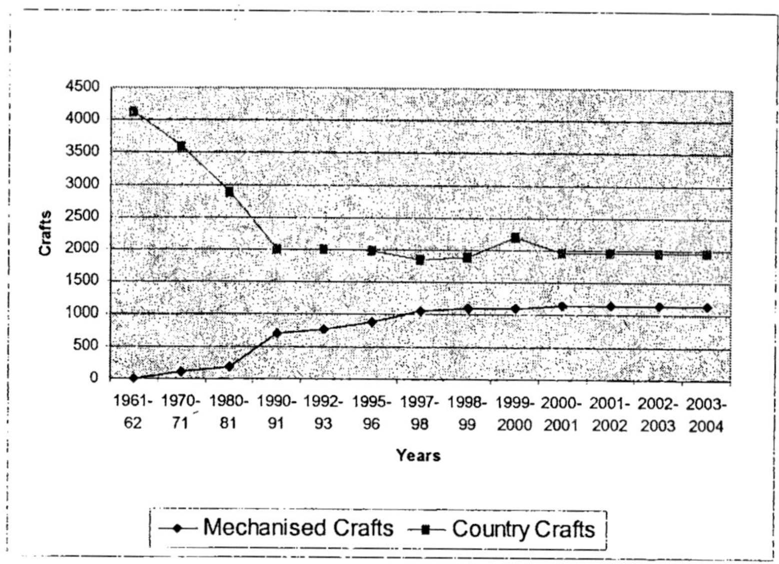

The deep-sea fishing and aquaculture were the other steps taken by the Government to increase production and export of fish. There are about 47 licensed farms of aquaculture covering an area of 200 hectares (Barbosa, Alexander M., Goa Today, and November 2002: 14).

These developments in the fishing sector have created adverse conditions to the traditional fishermen. It has led to the entry of non-fishermen into the fishing occupation with the sole motive of making maximum profit. It has led to the lopsided development of the fisheries as merchandised fishing received priority at the cost of the non-merchandised fishing. As per the recommendations of the Food and Agricultural Organization of the United Nations there should be 30 trawlers per 10 kms of coastline. Considering the Goa's coastline length of 105 kms, the number of trawlers should be around 315. But today, Goa has 1134 merchandised boats (Barbosa, Alexander M., Goa Today, November 2002: 12). This has endangered the sustainable exploitation of the marine resources.

Goa's annual fish catch includes a wide variety of shell-fish and fin-fish. The most popular finfish are the sardines and the mackerels, which constitute 50 per cent of the total fish exported to the outside markets like China, Japan, etc. It also constitutes 60 per cent of the total landing. In 2004, the landings of the other fishes such as catfish, sharks, seerfish, prawns, pomfrets, bombay ducks and butterfish were to the tune of more than 60,000 million tonnes. Most of it goes to the hotel industry. The shrimps and the pomfrets are exported and what is remaining is available for home consumption. Goa's demersal resources like shrimps, prawns, clams, oysters and mussels are located up to a distance of 200 meters from the shore.

The fishing industry not only markets fresh fish but also sun dried salted and canned fish. There is a good scope for setting up reservoir culture fisheries in Goa in its estuarine lowlands and mangrove vegetation (De Souza: 1990).

Goa's fish resources can be classified into two sectors namely; marine fisheries and inland fisheries. The major fish production comes from inshore fisheries, marine fisheries and off-shore fisheries. The inland fisheries include fishing with traditional method namely; the Rampons and Mags.

Inland Fishing

Goa offers good scope for aquaculture and fish farming. The tradition of brackish water fish culture is already age-old and well known. Low lying paddy fields known as khazan lands which cover an area of about 18,000 hectares have not been, so far, exploited in a planned and systematic manner for pisciculture (Angle : 2001 : 89).

Inland fish production of Goa comprises of both capture and culture yields. The type of fishing gears operated in the inland waters, particularly in the brackish waters are stake nets, nets operated at sluice gates, gill net, barrier net, cast net and small and medium sizes of seine nets. The fish catch consists of prawns, mullets, milk fish, sea bass, catfish, lady fish, pearl spot, croakers, carangids, snappers, etc. Goa has about 3,500 hectares of potential brackish water areas for fish culture, for which about 2000 hectares are fit for aquaculture. The brackish water resources are put at about 12,000 hectares of khazan lands in particular, with a potential to develop one crop of prawns, without sacrificing paddy cultivation.

There are 3,500 hectares of marshy land available in the state for being developed for brackish water prawn farming.

In addition to this, about 4000 hectares of low-lying land, which are utilized for paddy cultivation during the monsoon season are suitable for culture of prawns and fish for remaining part of the year. The fisheries potential is assessed to be around 70,000 tons per annum. As against this potential, the local catch is normally in the vicinity of 50,000 tonnes, which has increased from around 17,000 tonnes in 1961 to 4397 million tonnes in 2002. All other fish is locally consumed. On an average, about 2000 tonnes of prawns are exported every year with export earnings of Rs. 6 crore.

Fishing in Goa in the pre-liberation days was done in the traditional fashion. The post-liberation period witnessed the mechanization of fishing technically called the "Blue Revolution". Mechanized fishing fleet consisting of purse seines, trawlers and other fishing vessels of modern fishing gears came into the Goan waters. It had far reaching consequences. The fish production increased by leaps and bounds. It introduced mechanized crafts and crafts with outboard motors. In 2004, the annual fish production was estimated at about 88791 million tonnes of which marine fish landing was 84394 million tonnes and inland fish landing was 4397 million tonnes (Directorate of Fisheries, Panaji). A modern prawn hatchery is being set up. It is expected to supply adequate prawn seeds to encourage pisciculture in 4000 hectares of khazan lands.

The Department of Fisheries was set up in 1963 on the basis of the study carried on by the Central team of experts on the fishing potential of the Union Territory. Fishing activity was earlier looked after by the 'Development Office'. The fisheries department today is headed by a Director, with the necessary technical staff working under him. It surveys fishery activities and examines and explores their potential for marine resources, in inland, estuarine and sea areas, and renders assistance to fishermen in the matters of purchase of diesel engines, construction of hull and requisites like fish-nets, nylon and cotton twine, diesel oil, etc. It takes measures to improve the conditions of fishermen, with schemes of financial assistance in terms of loans of a medium or short-term nature, managerial subsidy, and supply of ice to them at concessional rates to enable them to preserve their fish catch. It maintains a laboratory for conducting biological research and a statistical wing, an offshore fishing station and a training centre. It also sends its own and fishing community's men for training at specialized centres of such learning in other parts of India. It provides for servicing and overhauling facilities to the mechanized fishing crafts. It also embarks on experiments in estuarine farming of fish and fosters in fishermen an interest in inland culture of fish, especially in marshy lands.

The Directorate of Fisheries has taken up a number of schemes to develop both marine and inland fisheries keeping in view the main objectives such as an increase in fish production, mechanization of fishing crafts, assisting export of fisheries products like processed canned fish and providing allied services for processing, preservation and marketing, berthing and landing facilities and giving practical training to fishermen in modern methods of fishing operations. These facilities help the fishermen in overcoming their marketing difficulties.

The Department of Fisheries also acts as prevention guard for illegal mechanized fishing along the Goa Coast. In order to enforce the provision of Marine Fishing Regulation Act, 1980, the Department has procured one patrol boat which is used in patrolling in specific areas along the coast. In order to safeguard the interests of the traditional fishermen along the Goa Coast regular patrolling is done to check any violation of the Marine Act. The Department of Fisheries has already taken steps to construct fresh water fish farms at Selauli and Anjunem to produce quality fish and to help private pisciculturists for propagation of fish culture. The Department has constructed jetties at Malim, Talpona and Cutbona. It has set up auction sheds, net building sheds, workshops, ice factories and cold storages at the landing centres. The Department also gives financial assistance in the form of loan and subsidy to purchase marine diesel engines, hulls, winches, purse-seines, webbings and other fishery requisites since liberation.

The Department has sanctioned Rs. 103.26 lakh as subsidy on kerosene and diesel, insulated boxes and acquisition of fishery requisites in the year 2004–05.

THE BLUE REVOLUTION

The Blue Revolution had its impact on Goa's economy and social fabric. Introduction of mechanized purse seines and trawlers, and motorization of traditional sailing crafts, has brought about a great deal of change in the practices of fishing in the state. The efforts to increase production had repercussions on the fish resources leading to the over exploitation and depletion of the fish resources, resulting in the conflict between the mechanized boat owners and the traditional fishing community. The mechanized trawlers in search of the valuable solar prawns began to fish very close to the shore, which are the spawning grounds for the variety of fish, particularly in the rainy season. The conflict between the traditional and mechanized fishing communities made the Government of Goa enact fishing regimes like the Goa Marine Act, 1980, the subsequent fishing rules and the ban on mechanized fishing within 5 km zone from the shore. Despite the fishing regimes, there has been over fishing. The fish crafts have increased over the years because of the liberal licensing policy of the Government. There are more trawlers and mechanized boats than necessary, thus, raising the issue of sustainability. The fish famine of 1980 forced the Government to fix the quota of fishing vessels. Now a N.O.C. is necessary for the construction of new boats. At the moment, the Government has fixed the construction of only 5 trawlers per year to take care of the replacement of old boats.

The population of fishermen is estimated at around 30,225. The effect of mechanization and motorization has contributed in great measures in attaining the target of 81,856 metric tones of marine fish production. In 1995, the Mormugao Port alone exported 13,474 tonnes of marine products mainly in the form of frozen shrimps fetching foreign exchange to the tune of around Rs. 67 crore.

Export and Tourism industry has given tremendous boost for fishing industry in Goa. For the projected population of 16 lakh of 2001-02, ICAR fixed the consumption to 16,000 million tonnes. The heavy investment in mechanized sector led to the achievement of a target of 1, 20,000 million tonnes. It was against the sanity of sustainable fishery growth.

The following factors have been responsible for the slow progress of fisheries sector:

1. Despite the monsoon ban indiscriminate fishing during the months of June and July continues. Juvenile

mackerels, which are caught, can neither be relished nor processed. They are dumped back into the sea leading to air and water pollution.

2. The mechanized trawlers in violation of the existing regulations operate mostly within the 5 fathom area and have hardly exploited areas beyond 10 fathoms. This has resulted in over fishing in shallow waters.

3. Trawling in shallow waters with the help of accessories like heavy weights and beams with a view to squeezing out prawns from the sea-bed slush has been responsible for destroying fish eggs. The intense trawling in limited areas has prevented natural biological regeneration of species.

4. Violation of regulation to the effect that the net mesh size should not be lower than 24 mm. It helps smaller fish to grow to its optimum size.

   It is strongly felt that Goa needs a fish farming technique which combines agriculture and aquaculture.

5. Increase in the number of mechanized boats beyond sustainable level (Angle, P.S.,2001:91).

MARINE FISHERIES (TRADITIONAL)

Near about 41 varieties of marine fishes are found in Goa. However, prominent among them are sardine and mackerel, which constitute almost 46 per cent of the oceanic catch (Verenekar, 2000). The commercial fish varieties like squid, cuttlefish, prawns, breams, ribbon fish, etc. contribute 58 per cent of the total marine export item.

The traditional sector has also taken up to Mechanization of their traditional crafts, which helps them in the netting operations. The Government has encouraged mechanization of the country crafts by outboard/ inboard motors by providing financial assistance in the form of subsidy to the tune of Rs.10,000 under Centrally Sponsored Scheme on 50:50 basis. Various schemes and programmes are being implemented in the state with the following objectives.

1. To increase the fish production by utilization of the available natural resources.

2. To ameliorate the socio-economic condition of the fishermen who belong to the weaker section of the society.

3. To upgrade the occupational training imparted to the fishermen in order to improve their operational skills and efficiency for tapping the resources of EEZ.

4. Strengthening of fisheries department by providing trained man power.

5. Production of fish seed and Reservoir Fisheries at Anjunem.

6. Integrated Coastal Aquaculture.

7. Demonstration cum Training in Brackish water Prawn/ Fish Farm.

8. Providing Landing and Berthing Facilities.

9. Enforcement and Protection of Reserved Fishing areas along Goa Coast.

10. Mechanization of Fishing Crafts by providing for subsidies on fishing vessels, outboard/ inward Motors etc.

11. Financial Assistance for construction of wooden/ F.R. Plastic Crafts.

12. Assistance for purchase of Fisheries Requisites.

13. National Welfare of Fishermen/ Group Accident Insurance Scheme for active Fishermen Education and Training in Fisheries.

14. Development of Inland Fisheries statistics.

It is pertinent to know the contribution of fishing industry in the Net State Domestic Product.

Table 2.3 shows the declining contribution of fishing industry to its Net State Domestic Product in recent years.

Table – 2.3- Contribution of Fishing Industry to Goa's Economy

Year	NSDP of Fishing Industry at Current Prices (Rs. lakh)	Total NSDP of Goa (Rs. lakh)	Share of Fishing Industry in NSDP(%)
1985-86	1100	50896	2.16
1986-87	1254	60787	2.06
1987-88	1138	67048	1.70
1988-89	906	78140	1.16
1989-90	1158	91995	1.26
1990-91	1272	98651	1.29
1991-92	1336	103226	1.29
1992-93	2625	145939	1.80
1993-94	3914	188651	2.07
1994-95	4752	209554	2.27
1995-96	6864	242877	2.83
1996-97	7584	257389	2.95
1997-98	8649	312760	2.77
1998-99	12470	521215	2.39
1999-00	11529	582741	1.98
2000-01	18386	663401	2.77
2001-02	15518	692551	2.30
2002-03*	13336	848592	1.57
2003-04*	11608	916024	1.27

* Provisional.

Source: Statistical Handbook and The Economic Survey, Government of Goa (Various Issues)

The contribution of the fishing industry to the development of the State economy may be measured in terms of:

1. Its contribution to Net State Domestic Product (NSDP) and

2. Its share in the total output of the primary sector.

Table 2.3 reveals that the contribution of fishing industry in Goa's NSDP since 1985 has never exceeded 3 per cent. There is a lack of steady growth in the contribution of fishing industry to the development of the Goa's economy. The decline in its contribution since 2001-2002 is a cause of concern.

Graph 2.2: Contribution of Fishing Industry to Net State Domestic Product

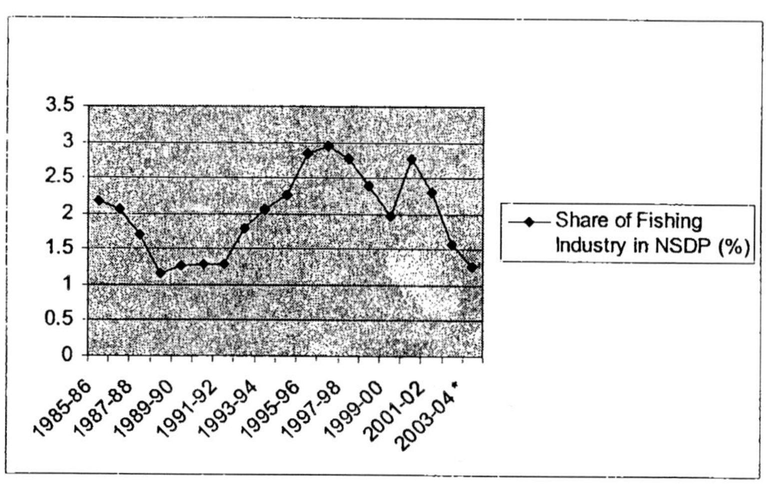

Table 2.4 depicts the share of the fishing industry to the primary sector. The contribution of the industry to the NSDP of primary sector has increased from 7.8 per cent at the time of statehood to higher double-digit figures in the last few years. A significant increase is evident in 2000-01 i.e. 18.65 per cent. But since 2001-02 it started falling.

Table – 2.4- Share of Fishing Industry in the Primary Sector

Year	NSDP of Fishing Industry at Current Prices (Rs. lakh)	NSDP of Primary Sector (Rs. lakh)	% Share of Fishing Industry in NSDP of Primary Sector
1985-86	1100	12374	8.89
1986-87	1254	14171	8.85
1987-88	1138	14543	7.83
1988-89	906	16744	5.41
1989-90	1158	18242	6.35
1990-91	1272	19846	6.41
1991-92	1336	21600	6.19
1992-93	2625	32714.5	8.02
1993-94	3914	43829	8.93
1994-95	4752	44911	10.58
1995-96	6864	47841	14.35
1996-97	7584	51963	14.60
1997-98	8649	54280	15.93
1998-99	12470	73872	16.88
1999-00	11529	78228	14.74
2000-01	18386	98567	18.65
2001-02	15518	94865	16.35
2002-03 *	13336	88021	15.15
2003-04 *	11608	100289	11.57

* Provisional.

Source: Statistical Handbook and the Economic Survey, Government of Goa (Various Issues)

Graph 2.3: Contribution of Fishing Industry to SGDP of the Primary Sector

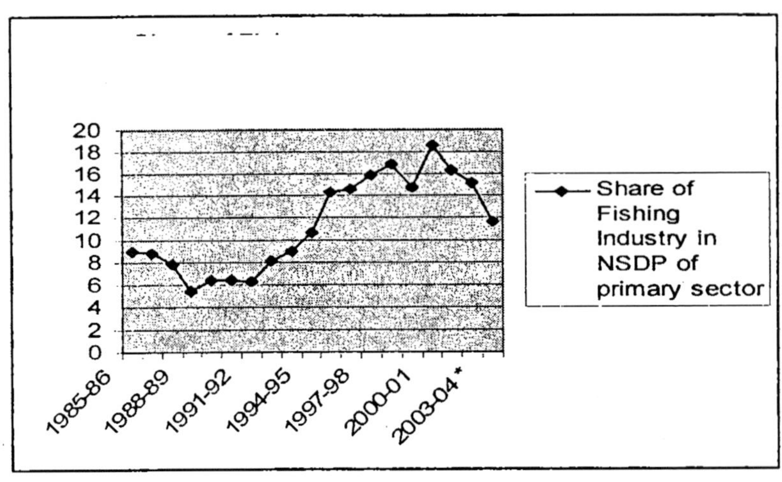

Table- 2.5- Number of Fishing Crafts and Gears Registered During 1961 to 2003

Sr. No.	Item	1961	1981	1991	2001	2003
1	2	3	4	5	6	7
1	Fishing Crafts					
	(a) Mechanised Boats	-	-	-	-	
	(i) Gill Netters (O.B.M.)	-	251	900	460	968
	(ii) Trawlers	-	282	-	255	1134
	(iii) Liners	4	36	837	-	
	(iv) Others	-	21	-	601	
	TOTAL	4	590	1757	1324	
	(b) Non-Mechanised Boats*	4125	2205	1900	1866	1700

1	2	3	4	5	6	7
2	Fishing Gears					
	Purse Seine				301	355
	(i) Drag nets	-	544	196	171	200
	(ii) Gill nets	-	4688	1925	2269	2450
	(iii) Trawl nets	-	580	598	1221	1500
	(iv) Cast nets	-	707	-	55	60
	(v) Trap	-	206	-	286	300
	(vi) Shore seines	-	424	-	93	-
	(vii) Spawn collecting nets	-	142	-	-	-
	(viii) Others	-	832	1342	842	695
	TOTAL	-	8123	4065	5238	5560

Source: Directorate of Fisheries, Panaji –Goa.

In the mechanized boat sector, the number of Gill-netters and Liners has increased tremendously from 1981 to 2003. The number of purse seine gears was 355 in 2003, which is quite high for the length of Goa's coastline. In the non-mechanized sector, the number has been reducing over the years from 1961.

In fishing gears, the number of Dragnets and Gill nets has reduced substantially. While the figures for trawl nets have seen a very slight increase, the number of other nets has increased considerably.

Table – 2.6 : Important Variety Wise Quantity of Marine Fish Catch (In M. Tonns) From 1991 To 2004

Year	Items							Total	
	Mackerels	Sardine	Catfish	Shark fish	Seer fish	Prawns	Pomfrets	Others	
1991	4305	28876	630	93	984	3231	242	37262	75623
1992	17062	14091	121	115	1673	2997	215	60059	96333
1993	20184	10175	500	1017	1010	3166	318	64552	100922
1994	38230	210	687	201	2016	4355	233	49908	95840
1995	42712	1653	1710	25	917	5193	1105	28541	81856
1996	44254	3461	151	501	859	4585	549	38377	92737
1997	21721	17120	684	964	1143	4174	1191	44280	91277
1998	19663	18530	650	583	1272	2651	524	23363	67236
1999	19240	21297	723	426	914	1622	371	15482	60075
2000	16589	19836	676	981	1398	2284	681	22118	64563
2001	14204	21470	2436	1211	2746	874	859	25586	69386
2002	8103	30951	1585	1355	1230	2299	500	21540	67563
2003	5779	30874	1007	1571	2274	6656	825	34770	83756
2004	6303	34203	1043	1305	3478	5586	568	31908	84394

Source: Directorate of Fisheries, Panaji – Goa

The data in Table 2.6 reveal the declining marine fish catch of most of the varieties. There has been a substantial decline in the catch of mackerels, shark, pomfrets, prawns etc. This decrease in fish catch is largely attributed to the exploitation of marine resources beyond sustainable limit.

Graph 2.4: Quantity of Mackerel catch in M.T. from 1991 to 2004.

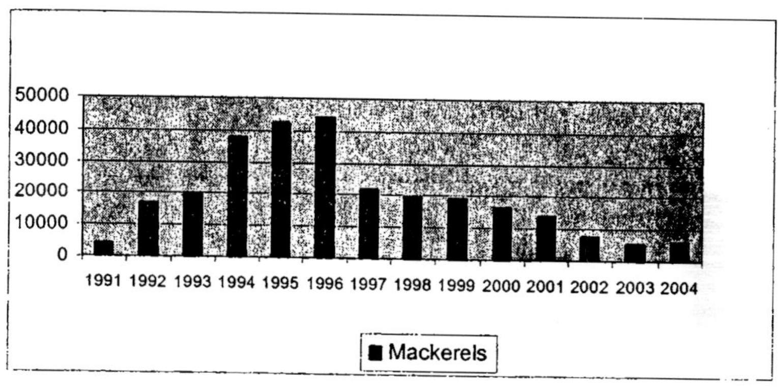

Graph 2.5: Quantity of Sardine catch in M.T. from 1991 to 2004.

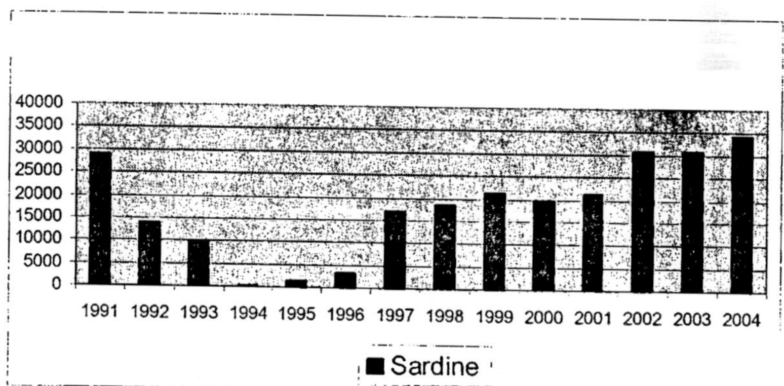

Graph 2.6: Quantity of Catfish catch in M.T. from 1991 to 2004.

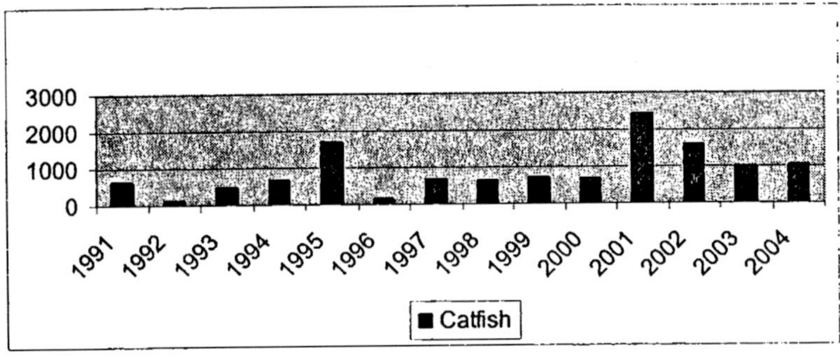

Graph 2.7: Quantity of Shark catch in M.T. from 1991 to 2004.

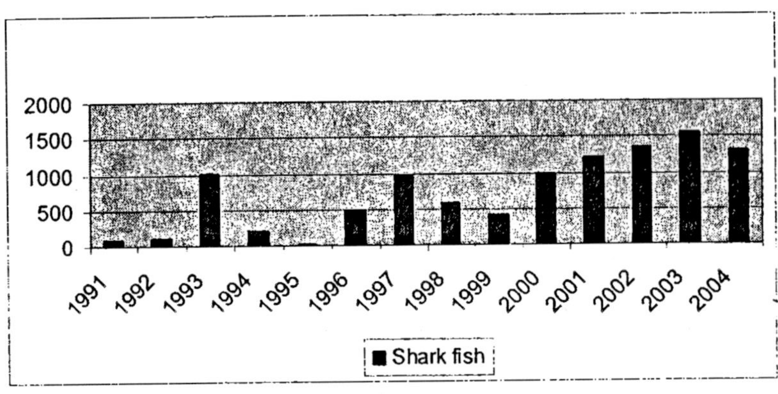

Graph 2.8: Quantity of Seer fish catch in M.T. from 1991 to 2004.

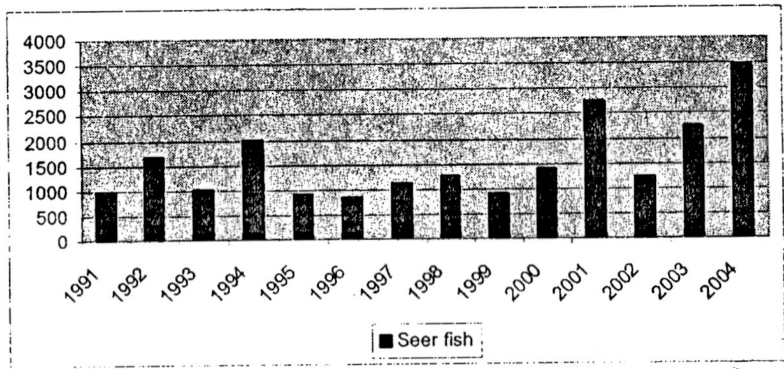

Graph 2.9: Quantity of Prawn catch in M.T. from 1991 to 2004.

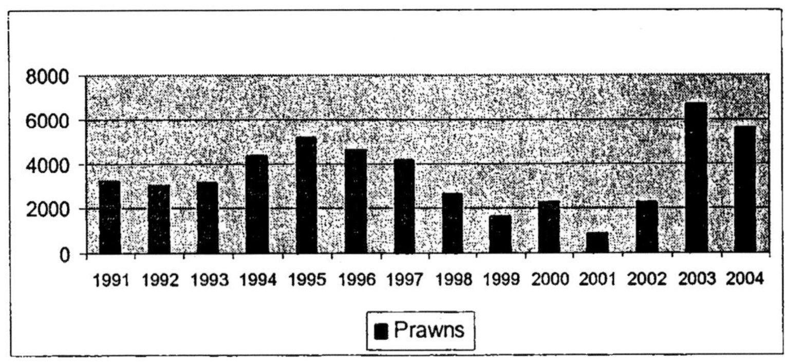

Graph 2.10: Quantity of Other fish catch in M.T. from 1991 to 2004.

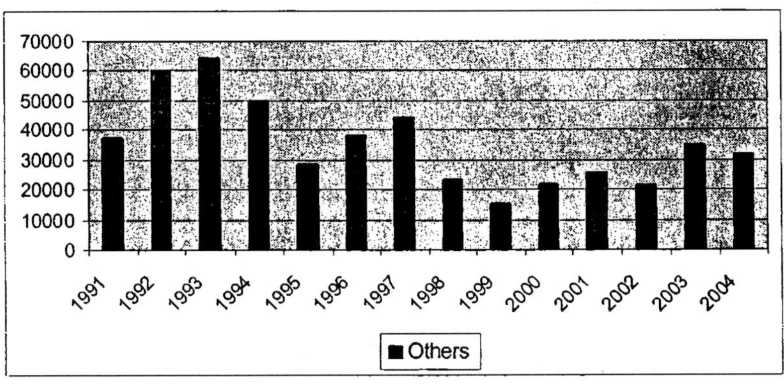

Graph 2.11: Quantity of Total Marine fish catch in M.T. from 1991 to 2004.

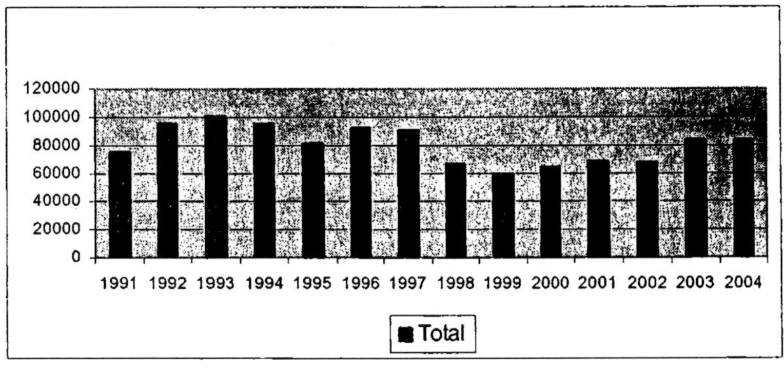

Table – 2.7 - Disposal of Marine Fish Catch for the Years 1971 to 2004

Sr. No.	Method of Disposal	Quantity in Tonnes					
		1971	1981	1991	2001	2002	2004
1	Marketed fresh	27586	26890	34457	54771	57669	75954
2	Sun dried	3598	890	4430	7493	2555	4220
3	Salted	2399	1314	2953	3506	5098	2532
4	Miscellaneous including manures	5997	2408	7384	3616	2241	1688
	Total	39580	31502	49224	69386	67563	84394

Source: Directorate of Fisheries, Panaji –Goa.

Graph 2.12: Quantity of fresh fish sold in local markets in M.T. from 1971 to 2004.

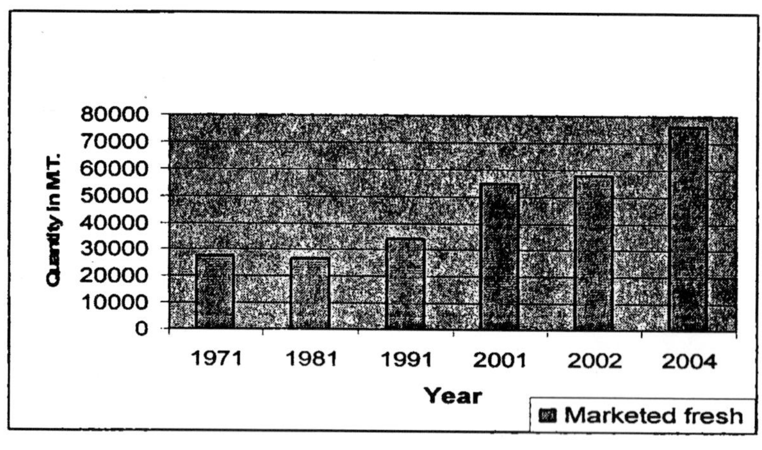

Graph 2.13: Quantity of sun dried fish sold in local markets in M.T. from 1971 to 2004.

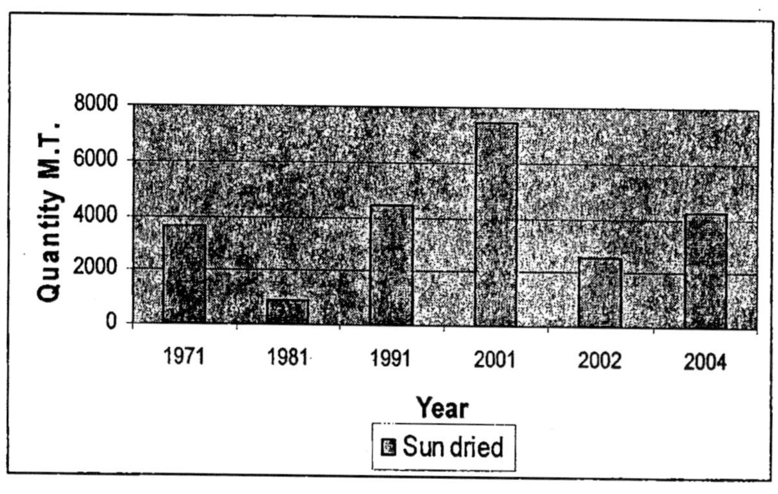

Graph 2.14: Quantity of salted fish sold in local markets in M.T. from 1971 to 2004.

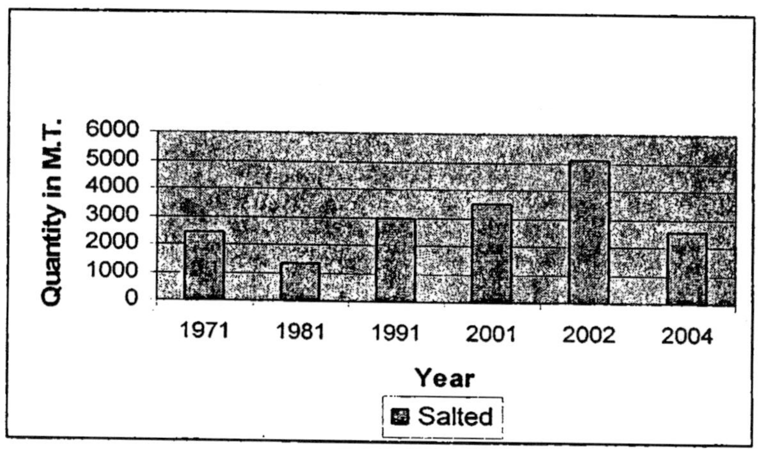

Graph 2.15: Quantity of fish sold for miscellaneous purposes in local markets in M.T. from 1971 to 2004.

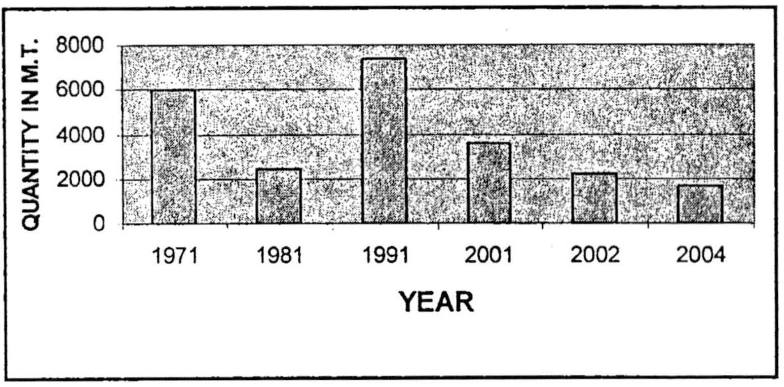

The quantity of all the variety of fish seen in Table 2.7 decreased in 1981 as compared to 1971 and then, further increased in quantity to more than 1971 quantity in 1991.

The figures from 1980-81 to 1999-00 shows that Goa's share in total fish production in India has been less than 2 per cent except for 3 years; 1992-93, 1993-94, 1994-95, where Goa's share marginally crossed 2 per cent.

Table- 2.8 - Fish Production in Goa and India during 1980-81 to 2001-02

Year	Goa*	India**	% share of Goa in total fish production in India
1	2	3	4
1980-81	0.26	24.42	1.05
1985-86	0.41	28.76	1.44
1990-91	0.56	38.36	1.47
1991-92	0.78	41.57	1.88
1992-93	0.99	43.65	2.27

1	2	3	4
1993-94	1.04	46.44	2.24
1994-95	0.99	47.89	2.07
1995-96	0.85	49.49	1.73
1996-97	0.96	53.48	1.80
1997-98	0.95	53.88	1.32
1998-99	0.71	53.62	1.18
1999-00	0.63	56.56 (p)	1.29
2000-01	0.68 (p)	—	—
2001-02	0.73 (p)	—	—

Source: * Directorate of Fisheries, Government of Goa

** Agricultural Statistics at a Glance, 2001, Directorate of Economics and Statistics, Ministry of Agriculture, Govt. of India,

P – Provisional

Graph 2.16: Percentage share of Goa total fish production in India from 1980 -81 to 1999- 00

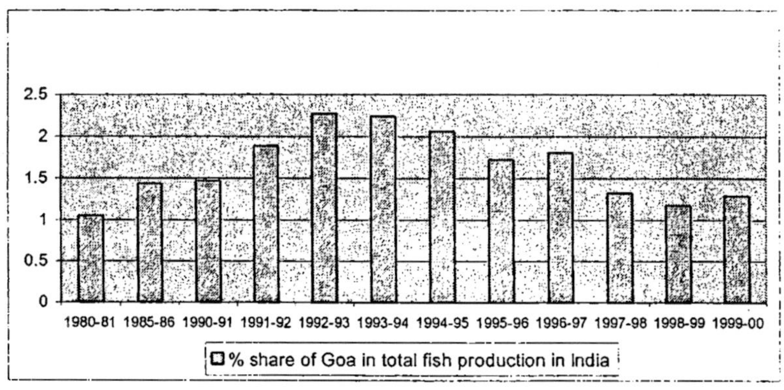

The percentage share of Goa in total fish production in India has also shown variation from decade to decade. The Goa's share in terms of percentage in the total fish production in India is also showing signs of decline.

CHAPTER III

Socio-Economic Conditions of Fishermen in Goa

The traditional characteristics of the fisher folk have undergone many changes though one finds continuity socio-economic characteristics. The conditions of life of the fishing communities vary not only from one country to other country but also within a country from one region to another region. Also, the extent to which they have undergone changes in these conditions differ substantially from region to region. In the developing countries, many fishing communities are the poorest and most neglected. Their status is low in the social hierarchy of a society. These communities have very less access to the formal education. Consequently, their representation in other occupations is either negligible or nil. Thus, job mobility is almost absent and there is no visible income from non-fishing activities. The basic services like water, electricity, roads, postal and telecommunications, shops, medical services, housing, banking institutions, etc., are not satisfactory. On account of small size of population and lack of leadership, they have failed to build up political pressure necessary to have favourable fisheries policies. They have very little political clout. In spite of all this, they had a deep knowledge of sea, fish and other resources available in their vicinity. They were highly skilled in their occupation.

However in the recent times the life of the fishing communities has undergone many changes factors like liberalization, globalization, mechanization, increased importance of fisheries sector as a source of nutrition and foreign exchange earner etc. are responsible for these changes. In India, the process of modernization of Indian fisheries started with the Indo-Norwegian Project. With the increased State patronage, the fisheries economy experienced a marked technological polarization. The introduction of trawlers and emergence of private mechanized sector were the other factors, which brought a sea of changes in the fishing sector. These factors have made deep impact on the socio-economic conditions of the fisher folk and have adversely affected the traditional fisher folk, who are relatively, socially and economically backward. The beneficiaries of these modern trends are the economically well to do fisher folk families as they were socially and economically in advantageous position to reap the benefits. The entry of non-fishermen in the fishing sector is another significant consequence of the process of modernization. These processes have not only increased the socio-economic gap between the different sections of the fishermen but have led to over exploitation and pollution of the sea.

SOCIO-ECONOMIC PROFILE OF FISHERMEN

In this chapter our main focus is to present the background factors pertaining to fisher folk. This not only provides a profile of the fisher folk but also an insight into the background factors that may be found relevant in the analysis of the issues under study.

Table- 3.1 - Age-wise and craft wise distribution of the respondents

Age	Traditional		Motorized		Mechanized		Total	
	No.	%	No.	%	No.	%	No.	%
Less than 21	1	1.1	1	0.7	-	-	2	0.5
21-30	9	9.7	33	23.1	25	15.2	67	16.7
31-40	25	26.9	27	18.9	60	36.6	112	28
41-50	22	23.7	34	23.8	38	23.3	94	23.5
51-60	22	23.7	39	27.3	33	20.1	94	23.5
61-70	13	14	7	4.9	6	3.7	26	6.5
71 and above	1	1.1	2	1.1	2	1.2	5	1.3
Total	93	100	143	100	164	100	400	100

The primary data reveals that in all the three categories the fishermen belonging to middle age group are in majority. The fishing activity is physically demanding. The long hours of sailing, loading of large quantity of fish etc requires the physical and mental strength to sustain the hard work. However, in recent years, the mechanization of fishing has reduced the strains and today one finds even aged fishermen engaged in fishing.

Table- 3.2 - Religion-wise distribution of the respondents

Religion	Traditional		Motorized		Mechanized		Total	
	No.	%	No.	%	No.	%	No.	%
Hindu	47	50.5	70	49	70	42.7	187	46.7
Christian	46	49.5	72	50.3	93	56.7	211	52.8
Muslim	-	-	1	0.7	1	0.6	2	0.5
Total	93	100	143	100	164	100	400	100

The religious composition of the fisher folk community indicates that the Hindus and Christians are the predominant constituents of this community. Both the Hindu and Christian population have the equal representation in our sample size. The Muslims have negligible presence in the fisher folk community.

Table – 3.3 - Caste-wise distribution of the respondents

Caste	Traditional		Motorized		Mechanized		Total	
	No.	%	No.	%	No.	%	No.	%
Kharvi	40	43.1	45	31.5	47	28.7	132	33
R. Catholic	47	50.5	62	43.4	94	57.3	203	50.7
Gabi	-	-	3	2.1	1	0.6	4	1
Bhandari	3	3.2	13	9	16	9.8	32	8
Maratha	3	3.2	20	14	5	3	28	7
Gold Smith	-	-	-	-	1	0.6	1	0.3
Total	93	100	143	100	164	100	400	100

The Hindu and Roman Catholic Kharvi are the significant communities found in the fishing activity. However, due to mechanization and consequent, high profit nature of the activity, non-traditional fisher folk communities are also getting attracted to this occupation. The mechanization and deep sea fishing require huge investment, which the traditional fishermen are unable to do. Therefore, other communities are entering into this business.

Table – 3.4 - Educational qualification of the respondents

Qualification	Traditional		Motorized		Mechanized		Total	
	No.	%	No.	%	No.	%	No.	%
Illiterate	38	40.9	33	23.1	18	11	89	22
Primary	19	20.4	33	23.1	24	14.6	76	19
Middle	16	17.2	26	18.2	36	22	78	20
Secondary	17	18.3	29	20.3	57	34.8	103	25
Hr. Sec	2	2.2	17	11.9	20	12.2	39	10
Graduate	1	1.1	5	3.5	9	5.5	15	4
Total	93	100	143	100	164	100	400	100

Graph 3.1: Educational qualification of the respondents.

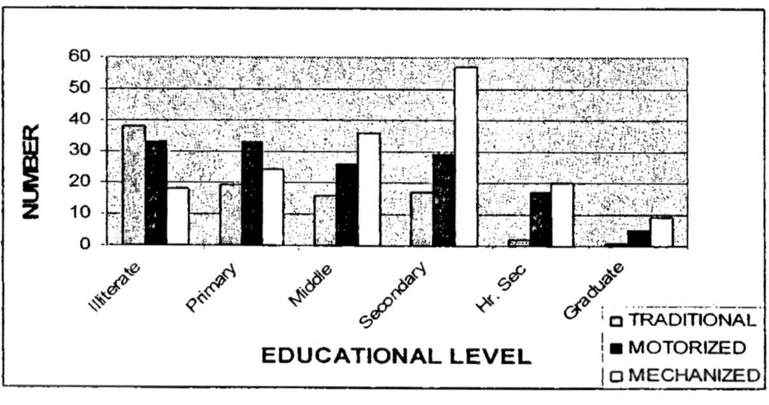

It is generally observed that the fisher folk communities are socially, educationally and economically backward. This is particularly true in case of the traditional fishermen. The field data clearly supports this view point. Illiteracy is widespread among the fisher folk. They are either illiterate or moderately literate. The percentage of people having education beyond secondary level is very low.

Table- 3.5 - Type of the family of the fishermen

Types of	Traditional		Motorized		Mechanized		Total	
family	No.	%	No.	%	No.	%	No.	%
Joint	30	32.3	47	32.9	48	29.3	125	31.3
Nuclear	63	67.7	96	67.1	116	70.7	275	68.7
Total	93	100	143	100	164	100	400	100

The primary data is in tune with the general trend of disintegration of joint family and emergence of nuclear family as a predominant institution of the contemporary society. About two-third of the respondents belong to the nuclear family. The effects of growing individualism, industrialization, urbanization, etc., seems to have also penetrated in the fisher folk communities.

Table – 3.6 - Size of the family of the respondents

Size of	Traditional		Motorized		Mechanized		Total	
family	No.	%	No.	%	No.	%	No.	%
1 - 2	5	5.4	2	1.4	2	1.2	9	2.3
3 - 4	33	35.4	42	29.3	63	38.4	138	34.5
5 - 6	29	31.2	67	47	79	48.1	175	43.7
7 - 8	11	11.8	27	18.8	13	8	51	12.7
9 - 10	6	6.5	3	2.1	5	3.1	14	3.5
11 - 12	7	7.5	1	0.7	1	0.6	9	2.3
13 - 14	2	2.2	1	0.7	1	0.6	4	1
Total	93	100	143	100	164	100	400	100

The common pattern of family being nuclear, majority of the fishermen families have 5-6 members. The small family norm seems to be well accepted by fisher folk in Goa. The bigger families are comparatively more among the traditional fisher folk than among the motorized and mechanized fisher folk communities.

Table – 3.7 - Attitude of the fishermen towards family planning

Family	Traditional		Motorized		Mechanized		Total	
Planning	No.	%	No.	%	No.	%	No.	%
In Favour	93	100	143	100	163	99.4	399	99.8
Against	-	-	-	-	1	0.6	1	0.2
Total	93	100	143	100	164	100	400	100

It is significant to note that almost all fishermen interviewed were in favour of family planning. It is due to this fact that the size of the family of the fisher folk is small. In general, the overall birth rate in Goa is less. It stands second in the country as far as low birth rate is concerned.

Table – 3.8 - Type of ownership of house

Ownership	Traditional		Motorized		Mechanized		Total	
of house	No.	%	No.	%	No.	%	No.	%
Own	93	100	140	97.9	162	98.8	395	98.7
Rented	-	-	3	2.1	2	1.2	5	1.3
Total	93	100	143	100	164	100	400	100

Almost all fishermen in Goa own the house. Very insignificant number of motorized and mechanized fishermen lives in rented houses. Ownership of house is one of the social indicators of standard of living of the people. The ownership of house by the large number of fishermen shows the better living standard of fishermen in Goa.

Table – 3.9 - Type of houses of the fishermen

Type of House	Traditional		Motorized		Mechanized		Total	
	No.	%	No.	%	No.	%	No.	%
Terraced	19	20.4	20	14	74	45.1	113	28.2
Tiled	72	77.4	123	86	90	54.9	285	71.3
Thatched	2	2.2	-	-	-	-	2	0.5
Total	93	100	143	100	164	100	400	100

It is not just enough to own the house. The type of house owned by the people is also important. The data reveals that majority of the fishermen possess tiled houses. A substantial number of them also own terraced house. It is important to note that about 45 per cent of the mechanized fishermen own terraced house, which is mainly due to good earning from the fishing business and the subsidiary income, mostly from the person working on a ship or abroad.

Table – 3.10 - Electrification of house

Response	Traditional		Motorized		Mechanized		Total	
	No.	%	No.	%	No.	%	No.	%
No	1	1.1	-	-	-	-	1	0.2
Yes	92	98.9	143	100	164	100	399	99.8
Total	93	100	143	100	164	100	400	100

The above data indicates the availability of electricity in all the houses of fishermen except one belonging to the traditional fishermen community. It may be useful to know here that in Goa 100 percent of the villages are electrified.

Table- 3.11 - Drinking water facility

Water facility	Traditional		Motorized		Mechanized		Total	
	No.	%	No.	%	No.	%	No.	%
Individual	74	79.6	107	74.8	157	95.7	338	84.5
Public tap	15	16.1	14	9.8	4	2.4	33	8.3
Well	4	4.3	22	15.4	3	1.8	29	7.2
Total	279	100	143	100	164	100	400	100

Availability of water is one of the aspects of infrastructure. The State of Goa receives an average rainfall over 3,000 mm. Due to this and for other reasons, the supply of water to both domestic and industrial purposes is comparatively good. The above primary data confirms this fact. About three-fourth of the households have their own water taps and other substantial number of fishermen draw water from the public taps. The well is another important source of water to some fishermen families.

Table- 3.12 - Toilet and bathing facilities for the fishermen

Facility	Traditional		Motorized		Mechanized		Total	
	No.	%	No.	%	No.	%	No.	%
Attached	68	73.1	94	65.7	155	94.5	317	79.3
Non-attached	25	26.9	49	34.3	9	5.5	83	20.7
Total	93	100	143	100	164	100	400	100

It is important to understand that all the fishermen have got toilet and bathroom facility. In all the three categories of fishermen, majority of them have the attached bathroom and toilet facility. The percentage of fishermen having these facilities is as high as 94.5 per cent in the mechanized category.

Table – 3.13 - Sanitation facilities

Sanitation	Traditional		Motorized		Mechanized		Total	
	No.	%	No.	%	No.	%	No.	%
Drainage	43	46.2	11	7.7	43	26	97	24
Sewage	-	-	12	8.4	5	3	17	4
None	50	53.8	120	83.9	116	71	286	72
Total	93	100	143	100	164	100	400	100

The absence of proper drainage and sewage system in the fishermen village is a cause of concern. About 72 per cent of the fishermen have neither drainage nor sewage facility. Most of the fishermen communities have their settlements near the sea where sand facilitates the sinking of water. Also, rearing of pigs by the fishermen provides a natural remedy to clean the traditional type of toilets.

Table- 3.14 - Availability of Government hospital facility to the fishermen

Distance in km	Traditional		Motorized		Mechanized		Total	
	No.	%	No.	%	No.	%	No.	%
1 - 5	47	50.5	118	82.5	106	64.4	271	67.7
6 - 10	35	37.5	25	17.5	57	35	117	29.3
11 - 15	11	12	-	-	1	0.6	12	3
Total	93	100	143	100	164	100	400	100

Availability of health facilities is another indicator of social development. About 68 percent of the fishermen belonging to all the three categories have the government hospitals within a distance of 1-5 km from their residence. For another 29 per cent the government hospitals are available within a distance of 6-10 km. Very few of them have to travel 11-15 kms to avail the government hospital services.

Table- 3.15 - Availability of Private hospital facility to the fishermen

Distance in km	Traditional		Motorized		Mechanized		Total	
	No.	%	No.	%	No.	%	No.	%
1 - 5	38	40.6	69	48.3	109	66.5	216	54
6 - 10	45	48.4	50	35	53	32.3	148	37
11 - 15	10	11	24	16.7	2	1.2	36	9
Total	93	100	143	100	164	100	400	100

Apart from government hospitals, the private clinics/ hospitals are also available to the fisher folk at a not much distance. More than 50 per cent of the fisher folk have private clinic/hospitals within a distance of 5 kms. Another 27 per cent have to travel about 6-10 kms to get the services of private doctor. Only about 9 per cent have to travel more than 10 kms for such facilities and most of them belong to the traditional and motorized categories.

Table – 3.16 - Possession of household assets

Assets	Traditional		Motorized		Mechanized		Total	
	No.	%	No.	%	No.	%	No.	%
Sofa	47	50.5	57	39.9	152	92.7	256	64
Gas stove	90	96.8	128	89.5	164	100	382	96
Cooker	80	86	112	78.3	162	98.8	354	89
Radio	77	83.8	120	83.9	161	98.2	358	90
Sewing machine	40	43	74	51.7	101	61.6	215	54
Refrigerator	64	68.8	108	75.5	159	97	331	83
Television	89	95.7	128	89.5	164	100	381	95
Motor bike	64	68.8	106	74.1	153	93.3	323	81
Car	6	6.5	5	3.5	80	48.8	91	23
Pick-up	2	2.2	2	1.4	43	26.2	47	12
Telephone	48	51.6	73	51	148	90.2	269	67
Mobile	14	15.1	8	5.6	78	47.6	100	25

The possession of household assets is another significant indicator of standard of living of the people. Our field survey reveals that majority of the fishermen possess not only basic household assets but also other items. Most of them have the household assets like sofa, gas stove, cooker, radio, sewing machine, refrigerator, television, motorbike, etc. It is evident from the data that the high percentage of people owning sofa, refrigerator, motorbike, car, pick-up, telephone and mobile belong to the mechanized sector. The possession of these household assets by large number of fishermen indicates the comparatively good standard of living of the fisher folk in Goa.

Table – 3.17 - Land ownership

Land owned	Traditional		Motorized		Mechanized		Total	
(in acres)	No.	%	No.	%	No.	%	No.	%
1	2	2.2	7	4.9	3	1.8	12	3
2	2	2.2	4	2.8	2	1.2	8	2
3	-	-	1	0.7	-	-	1	0.2
4	-	-	-	-	1	0.6	1	0.2
No land	89	95.7	131	91.6	158	96.3	378	94.6
Total	93	100.1	143	100	164	99.9	400	100

The pattern of distribution of land among the people is one of the factors revealing how well the society is equalitarian in nature. It is evident from the data that the fishermen are deprived of land ownership. It means the fishing is the main occupation and source of income to the fishermen in Goa. It is only about 5 per cent of the fishermen are the owners of land, mostly just about 1 or 2 acres.

Table – 3.18 - Number of fishermen obtaining LIC Policies

Response	Traditional		Motorized		Mechanized		Total	
	No.	%	No.	%	No.	%	No.	%
No	66	71	117	81.8	72	43.9	255	63.7
Yes	27	29	26	18.2	92	56.1	145	36.3
Total	93	100	143	100	164	100	400	100

It is surprising to note that a large number of fishermen have neither insured their life nor the boat and other equipments/ instruments. It is primarily attributed to ignorance and lack of understanding of the significance of insurance. It brings out the need to create awareness among the fishermen about the utility of insurance policy. As high as 71 per- cent of the traditional fishermen and about 82 per cent of the motorized fishermen have not made the LIC policies. Among those who have taken LIC policy, majority (56 per cent) belong to the mechanized sector.

Table – 3.19 - Reasons for accepting fishing as occupation

Label	Traditional		Motorized		Mechanized		Total	
	No.	%	No.	%	No.	%	No.	%
No alter-native job	11	11.8	19	13.3	18	10.9	48	12
Fishing is profitable	5	5.3	3	2.1	26	16	34	8.5
Family occupation	3	3.2	1	0.7	5	3	9	2.3
To help father	-	-	1	0.7	2	1.2	3	0.7
Interested in fishing	70	75.3	119	83.2	108	65.9	297	74.2
Lack of employment	2	2.2	-	-	3	1.8	5	1.3
Parents force	2	2.2	-	-	2	1.2	4	1
Total	93	100	143	100	164	100	400	100

The fishermen have given varied reasons for taking to fishing occupation. The most dominant reason appears to be interest in fishing. However, it should not be literally interpreted as keen interest in pursuing the fishing occupation but more as a traditional/ family occupation. The fishermen are engaged in fishing from generation after generation. It is rather their hereditary occupation.

The children observe, learn and involve in fishing activity from their early age. This creates in them interest in their family occupation. Some continue to engage in fishing due to non-availability of alternative jobs or lack of employment. A profitable nature of occupation draws some people to this occupation.

Table – 3.20 - Family members involved in allied activities

Label	Traditional		Motorized		Mechanized		Total	
	No.	%	No.	%	No.	%	No.	%
Sale	11	12	1	0.7	3	-	15	4
Net making	2	2.2	-	-	-	-	2	0.5
Boat Construction	1	1.1	-	-	-	-	1	0.2
Personal	-	-	2	1.4	20	12.2	22	6

The fishing occupation creates many allied jobs like pickle-making, net making, boat construction, sale of fish etc. The data indicates that these allied jobs in which substantial number of fisher folk were engaged earlier are on the verge of extinction. This is due to the availability of ready made things in the urban markets. Very few are now engaged in such jobs. The mechanization, deep sea fishing etc have displaced the people from such jobs.

Table – 3.21 - Distance of fishing area from district headquarters

Distance	Traditional		Motorized		Mechanized		Total	
	No.	%	No.	%	No.	%	No.	%
1 - 5	3	3.2	16	11.2	63	38.4	82	20.5
6 - 10	21	22.5	1	0.7	1	0.6	23	5.7
11 - 15	3	3.2	121	84.6	32	19.5	156	39
16 - 20	52	56	5	3.5	44	26.9	101	25.3
21 - 25	14	15.1	-	-	24	14.6	38	9.5
Total	93	100	143	100	164	100	400	100

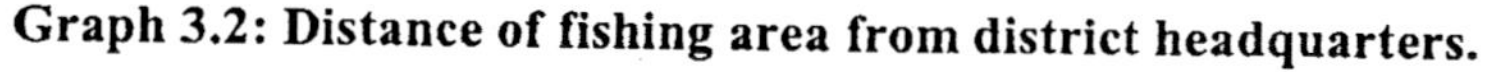

Graph 3.2: Distance of fishing area from district headquarters.

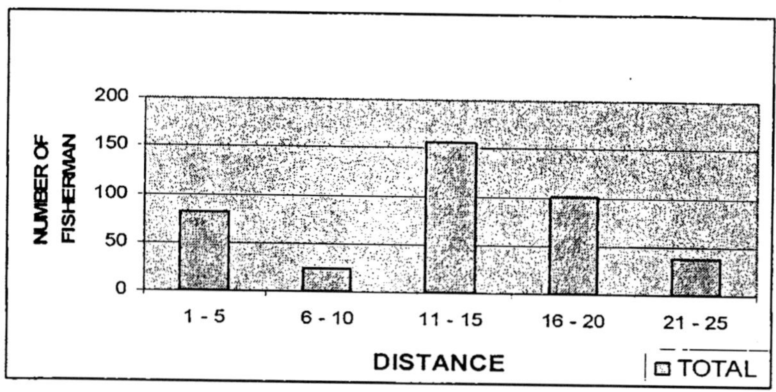

Distance of fishing area from the place of residence and market is important in the context of efficiency, time, marketing, etc. Thc fishing activity has to establish relationship with the district headquarters for various official works. Majority of the fishermen have to travel a distance ranging from 11-25 km to reach district place for any official work related to fishing, which means more expenditure and time consuming.

Table – 3.22 - Distance of fishing area from Main Road

Label	Traditional		Motorized		Mechanized		Total	
	No.	%	No.	%	No.	%	No.	%
1 – 5	17	18.2	95	66.4	63	38.4	175	43.7
6 – 10	10	10.8	1	0.7	1	0.6	12	3
11 – 15	15	16	42	29.4	56	34	113	28.3
16 – 20	51	55	5	3.5	44	27	100	25
Total	93	100	143	100	164	100	400	100

Graph 3.3: Distance of fishing area from main road.

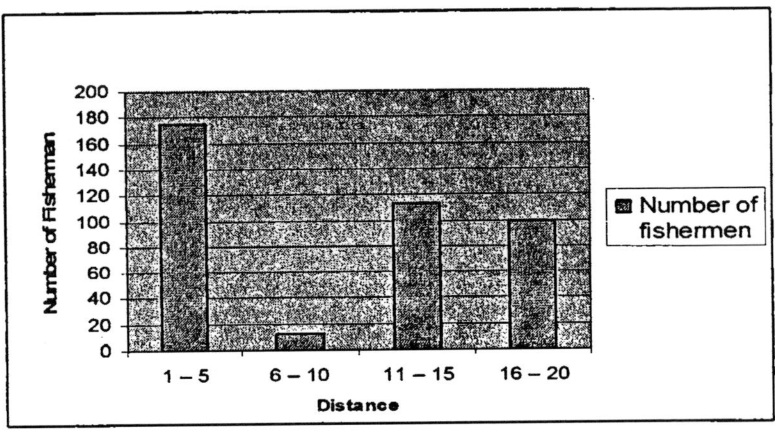

The data regarding distance of fishing area from main road reveals that a good number (175) of fishermen have their fishing area within 1-5 km from the main road. However, for more than 50 per cent fishermen this distance is between 11-20 km. Easy access to main road is very important from the point of view of transporting the fish to the market.

Table- 3.23 - Distance of fishing area from the near market

Label	Traditional		Motorized		Mechanized		Total	
	No.	%	No.	%	No.	%	No.	%
1 - 5	44	47.3	99	69.2	70	42.7	213	53.3
6 - 10	10	10.7	1	0.7	3	1.8	14	3.5
11 - 15	14	15.1	43	30.1	55	33.5	112	28
16 - 20	25	26.9	-	-	36	22	61	15.2
Total	93	100	143	100	164	100	400	100

Graph 3.4: Distance of Fishing Area from market

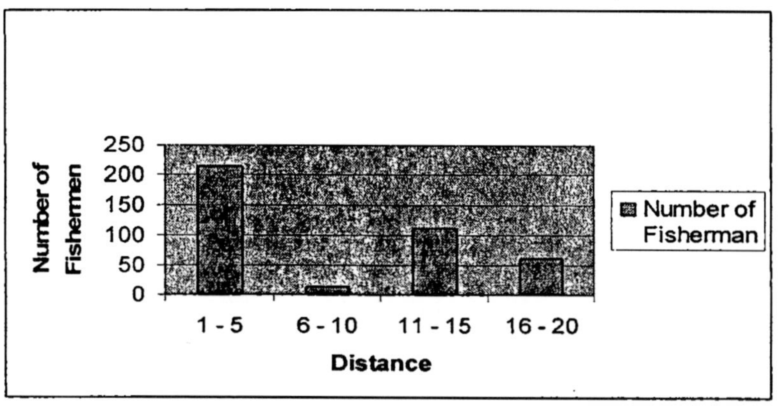

Distance of fishing area from the market is also significant from the view point of marketing. The less distance not only provides easy access to market but it also reduces the cost of transportation of fish and time consumption. A field data in this regard suggests that majority of the fishermen have market within a distance of 5 kms. However, more than 43 per cent of the fishermen have to travel a distance between 11-20 kms to reach the nearby market.

The general perception of the people about fishing communities is that they are economically backward, educationally poor, occupying low social status and culturally and politically isolated. The fishing communities have representation of people belonging to all religions. It is generally held that the people in the fishing sectors are some of the poorest and most neglected. In many states of India the fisher folk have remained socially, educationally and economically backward. The standard of living is low.

Our primary data collected through fieldwork contradicts the above conditions of the fisher folk community. The socio-economic conditions of the fishing communities in Goa appears to be different to a great extent. A large majority of the fishermen have their own houses with essential facilities like electricity, drinking water, separate bathroom and toilet etc. Moreover, the fishermen households possess both essential and luxury assets like gas stove, cooker, sofa, radio, Refrigerator, T.V., etc., which certainly indicate the better living standards of life. Educationally, too, they are not very backward. In recent years, they are taking to education. The schools and colleges are easily accessible to them. They are also politically active. The health facilities are available at a shorter distance. However, there is variation in the living conditions of the fishermen belonging to traditional, motorized and mechanized categories. The mechanized fishermen are more advanced economically and socially than the traditional fishermen. They have more easy access to education, health services and facilities like car, phone etc. With the promotion of mechanization and deep sea fishing by the government, many non- traditional fishermen are also entering into the fishing business. These people have the necessary capital and other resources required to undertake mechanized fishing. One important contributory factor for better living standards of fishermen in Goa is the fact that many youth particularly from Roman Catholic religion are working in ships earning attractive salary.

CHAPTER IV

ROLE OF WOMEN IN FISHERY SECTOR

The role of men and women in fish production, processing and marketing differ but one cannot deny the fact that women are playing significant role in the fisheries sector. In most of the countries of the world, women have gained the knowledge to manage the small boats, to make and mend nets, to process and sell fish. It is a known fact that women are playing leading role in the development of aquaculture. After the fish is brought to the shore, the women generally do most of the fish processing, preserving, storage and also selling the fish at the market (S. Suwanrangsi, 1998). However, in spite of all this, women's role in fisheries is undervalued. The type of role the women can play and the extent to which they can play those roles is largely dependent upon the socio-cultural milieu of the society. It is for this reason that the United Nations declaration of the Decade for women in 1975, paved way for the initiation of several measures "to raise the profile of the age-old social, cultural and economic barriers that prevented and hindered women from being active agents and beneficiaries of development" (K. Rana and V.Q.Perrez-Corral, 1998). In this context, promotion of maximum participation and empowerment of women in the fisheries sector is essential. The sustained growth of productivity, and sustainable use of fisheries and other natural resources is impossible without the recognition of the crucial role of the women. Women will have to be equal partners and effective participants in fisheries activities in order to improve nutritional and living standards both of themselves and their families. They should get access to new and appropriate technologies for their effective participation in sustained fisheries development. Thus, increased participation of women in various fisheries activities and also in the decision-making is essential. Because "despite the major roles of women in the sector, most women lack access to and control over physical and capital resources to decision-making and leadership positions, and to training and formal education. Access to these critical resources and services would improve the efficiency, profitability and sustainability of their activities. In the process, women would be empowered to become active agents in economic and social changes that will uplift their lives and those of their families" (Ibid, 1998).

In this chapter an effort is made to understand the role of women in fisheries sector in the context of Goa. In Goa, the socio-cultural milieu is conducive to promote the full participation and empowerment of women in the fisheries sector particularly in the contemporary times. Social customs in Goa do not prohibit the women from engaging in the supply and vending of fish in local markets. Given the opportunities, women in Goa have the potential to become important fish entrepreneurs.

As in other parts of the country, women in Goa never participate in the task of fish catching. Their main role is in marketing. In the present study, role of fisherwomen is analyzed in the context of marketing the fish and not in salting, drying and other processing activities. Unfortunately, there is no statistics on the number of women working in the fish markets or fish processing plants. The role of women as facilitators of fish distribution is indeed significant. Although women retailers form only a small segment of the total fish trade, they are a vital link between wholesalers and consumers (Bhatta: 2003).

Rubinoff (1999) has studied the impact of development on Goa's fisherwomen. According to her, there are hundreds of women who work daily in the four largest town markets and approximately 20 per cent of these could be considered "entrepreneurs" in their business. There are hundreds of other fisherwomen, some who work only part-time, who peddle fish in numerous villages and smaller towns or sell fish by the side of the road near bus-stops, bridge crossings, on the foot-paths near the town markets, etc. Until the Indian and State Governments' emphasis on the development of a modern fisheries industry in Goa during the mid 1970's, those involved in fishing activities were mainly people from the indigenous Kharvi or Gabi castes (Rubinoff: 1999).

But today the situation has changed to a great extent partly due to the expanding complexity of the fisheries industry in which a broad range of non-fishing groups sought employment and partly due to growing influx of migrant workers from neighbouring states that also enter into fish selling business along with other activities. Besides, there have been increased educational and occupational opportunities for lower socio-economic strata like the fishing community. Within the fishing population also many structural changes have taken place. Many Kharvi youth are migrating to the Gulf countries or Europe for better work opportunities and are moving into alternative, non-fishing occupation, such as construction or hotel service. Remittances sent home to families have brought about a positive impact on the scenario of fishing villages more prominently in housing development.

For the study of working conditions of fisherwomen, interviews were held with the 15 fisherwomen each in Margao, Vasco, Panaji and Mapusa town markets. In all 60 fisherwomen were interviewed. Their marketing tricks were also observed for this study. It is a well-known fact that there is increased migration of unemployed youth and poor communities attracted by higher wages in Goa from the neighbouring states. This migration has considerably increased the number of "outsiders" (both non-Goan and non-Kharvi people) involved in fishery- related activities, including transportation and fish processing, ice and netmaking industries.

An in-depth discussion with the fisherwomen revealed that men from Kharvi community do not sell fish in the local markets though in recent years we see some men selling fish in the town markets. One finds that majority of the fish-vendors are women.

With the growing mechanization, some men have moved into marketing niches at the jetties as middlemen, auctioneers or bankers. Often the husbands and the brothers-in-law of the fisherwomen purchase fish by auction at the jetties and give to these female vendors to sell in the markets of their destination. Sometimes, women themselves bargain directly at the jetty and either purchase fish for their own business or resell to other women in the daily markets.

THE SOCIO-ECONOMIC PROFILE OF THE FISHERWOMEN

Table- 4.1 Age Composition of the Respondents

Sr. No.	Age (in yrs)	Number of Respondents	Per cent
1	21-30	2	3.3
2	31-40	26	43.3
3	41-50	25	41.7
4	51-60	7	11.7
	Total	60	100

As is evident from the above data, majority of the respondents belong to the adult age. It is generally observed that the women belonging to adolescent and old age are not seen among the fisherwomen. It could be due to the factors like old age in case of women above 60 years and the lack of marketing skills, immaturity, unwillingness to expose adolescents to the public etc in the case of adolescents. About 85 per cent of the fisherwomen belong to the age group 31-50. This indicates the predominance of middle aged women among the fisherwomen as the marketing of fish is both physically and mentally demanding.

Table - 4.2 Educational Qualifications of Respondents

Sr. No.	Educational Qualification	Number	Per cent
1	Illiterate	52	86.6
2	Primary	7	11.7
3	Middle	1	1.7
	Total	60	100

One can find widespread illiteracy among the fisherwomen. The marketing of fish is a traditional occupation requiring no formal education. This means education has no much relevance to the type of job these women handle. The simple arithmetic is enough to sell the fish, which is learnt by practice. However, in recent years the female children of the fisher folk community are taking to education.

Table - 4.3 Husbands' Occupation

Sr. No.	Husbands' Occupation	Number	Per cent
1	Vegetable seller	1	1.7
2	Watchmen	3	5
3	Toddy tapper	4	6.6
4	Fishermen	18	30
5	Industrial worker	1	1.7
6	Driver	1	1.7
7	Casual labour	6	10
8	Farmer	3	5
9	Shoe maker	1	1.7
10	Carpenter	2	3.3
11	Truck driver	1	1.7
12	Govt. servant	1	1.7
13	Coolie	1	1.7
14	Not working	17	28.2
	Total	60	100

Generally, the fisherwomen belong to the fisher families. However, with increased industrialization and availability of other avenues, the members of the fisher folk community are taking to other occupations. The fishing activity is becoming unattractive, risky and non-remunerative to the youth. Therefore, they are taking to other jobs. It should be noted that the women belonging to non-fishing family are also getting involved in selling of fish. It is important to note that significant number of husbands of fisher women are not working for reasons like old age, alcoholism, etc. In such families, fisher women are the only earning members.

Table - 4.4 Number of years involved in fish sale

Sr. No.	Number of years	Number	Per cent
1	1-10	20	33.4
2	11-20	27	45
3	21-30	10	16.6
4	31-40	3	5
	Total	60	100

Graph 4.1: Number of years involved in fish sale

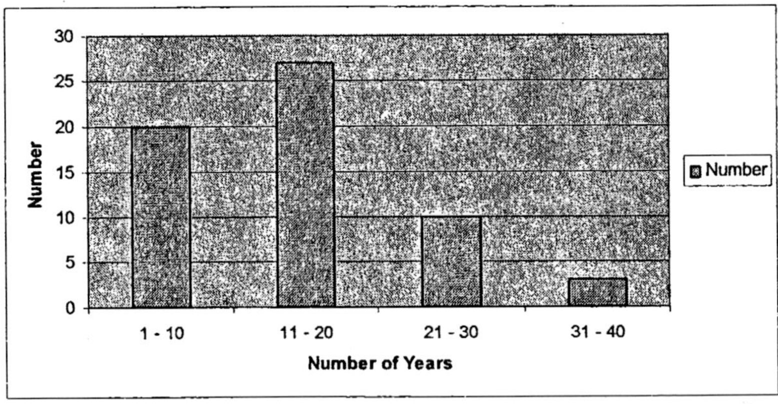

It is evident from the data that the women are involved in the sale of fish for a long time. More than 50 per cent of the women are in this occupation for more than 20 years. As this is generally a family occupation they take to selling of fish at an early age. Most of the fisherwomen are married.

Table- 4.5 Sources of fish bought for retail selling

Sr. No.	Source	Number	Per cent
1	None	12	20
2	From market	21	35
3	Other fisherwomen	15	25
4	Fishermen	3	5
5	Fish dealer	6	10
6	Agent	3	5
	Total	60	100

Graph 4.2: Sources of Fish bought for retail selling

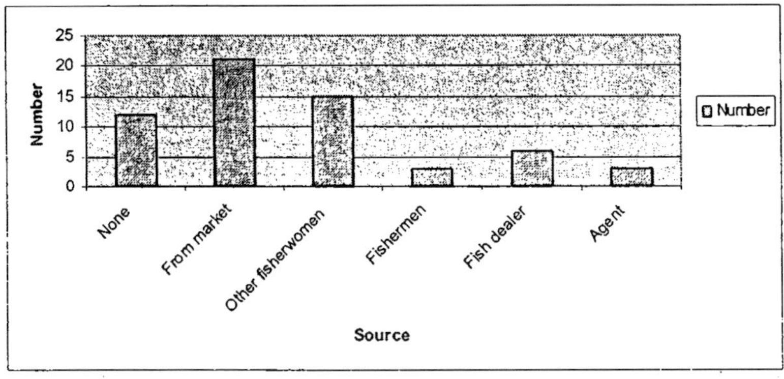

Fisherwomen purchase fish for retail sale from various sources. It is significant to note that most of the fisherwomen buy fish either from the wholesale market or from other fisherwomen, fish dealer, etc. However, this cannot be construed that the fisherwomen do not belong to fishermen families. They buy fish from various sources for the reasons like less fish catch, failure to get fish through the fishing activity, abandoning of fishing activity for some reasons etc. However, in recent years women from non-fishing families are also involving in fish selling business. The work of the fisherwomen starts very early in the morning. Everyday in the morning the women anxiously wait for the return of boats and trawlers. After their return, they participate in a competitive auction, which operates within a broad frame work of social hierarchy. Whenever the catch is less, the auctioning acquires greater significance because lesser the fish greater will be the demand. In such a situation the person who can make the highest bid will purchase and later on can resell for a good price. Those fisher women with higher financial base are the beneficiaries. If there is more catch even the poorer sections can participate in the market activities effectively. Over the years a class hierarchy has been formed among the fisherwomen. On the top of the hierarchy are found those fisherwomen who either have their own men-folk on board of a boat or trawler or they are capable enough to purchase the fish directly from the trawlers and boats and resell the same to other ladies. In the middle of the hierarchy are found those women who purchase the fish from the first category of women and sell it in the market. In the last echelon are situated women who purchase small shares of fish from various sources.

Table - 4.6 Quantum of fish bought for sale

Sr. No.	Source	Number	Per cent
1	500	6	10
2	700	5	8.3
3	800	3	5
4	1000	18	30
5	1500	4	6.7
6	2000	20	33.3
7	3000	3	5
8	6000	1	1.7
	Total	60	100

Graph 4.3: Quantum of fish bought for sale.

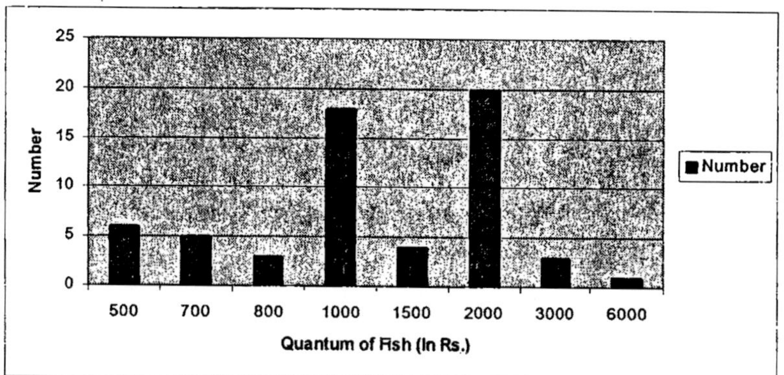

Large number of fisherwomen buy fish for retail sale ranging from Rs. 500 to Rs. 2000. It means, most of the fisherwomen are small-scale fish vendors. The fish being a perishable item, the fisherwomen buy fish in small quantity enough for a day's sale.

Table- 4.7 Mode of Transport

Sr. No.	Mode of Transport	Number	Per cent
1	Bus	22	36.7
2	Rickshaw	32	53.3
3	Tempo	5	8.3
4	Ferry	1	1.7
	Total	60	100

The rickshaw appears to be a predominant mode of transport of fish to the fish market. The public buses are the second important mode of transport to the fisherwomen to take fish to the market for sale and usually they travel by the same bus everyday. Tempo is made use of in case there are many baskets. If the distance between the places where they get fish and the market yard is not much, the women carry the fish on the head.

The primary data reveals that the amount spent on transport by majority of the fisherwomen varies from Rs. 10 to Rs. 40. Very few spend above Rs. 40 on transport. Apart from the expenditure on transport, the fisherwomen also incur expenditure on ice, food, etc.

The bamboo basket is more widely used to transport and sell fish in the market. However, in recent years, plastic vessels are also being used by some fisherwomen. The use of ice is the common form to preserve the fish. However, salt is also used by some to protect the fish from getting spoiled. However, in spite of this the wastage problem prevails due to lack of cold storage facility in the rural areas.

Table- 4.8 Average time taken for the sale of fish

Sr. No.	Time	Number	Per cent
1	6 hours	7	11.7
2	8 hours	15	25
3	9 hours	2	3.3
4	10 hours	21	35
5	12 hours	15	25
	Total	60	100

Graph 4.4: Average time taken for the sale of fish

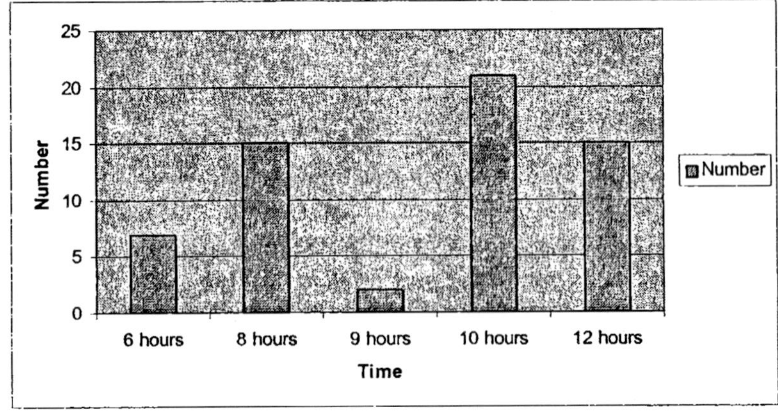

The fisherwomen spend many hours on selling the fish. A vast majority of them spend 8 – 12 hours per day on procuring and selling the fish.

Table- 4.9 Problems faced by the women in selling the fish

Sr. No.	Problems	Number	Per cent
1	No problems	35	58.2
2	Space problem	4	6.7
3	Customer problem	4	6.7
4	Dustbin problem	4	6.7
5	Pay heavy sopo	1	1.7
6	MMC problem	5	8.3
7	No profit	1	1.7
8	Transport problem	1	1.7
9	Other fisherwomen problem	2	3.3
10	Paying heavy electricity bills	3	5
	Total	60	100

The fisherwomen face several problems both in procuring and selling the fish. Shortage of fish, high rates and transportation are some important problems faced in procuring the fish. The important problems faced at the place of market are; non-availability of sufficient and proper space, payment of heavy tax, tough competition, customer problems like too much bargaining, fights with other fisherwomen, lack of basic facilities like water, dustbin, toilet etc at the market place, unviable profit margin and so on. Lack of shelter and harassment from the officials were other problems faced by them. They also suffer from several health problems. The process of conducting their trade itself is a problem for various reasons. First of all, for the whole day they have to continuously shout about their respective varieties of fish and the prices. Many women feel that the shouting takes much of their energy everyday. Secondly, they have to deal with the bargaining of every consumer which is irritating and time consuming.

Table- 4.10 Domestic problem faced by the women

Sr. No.	Problems	Number	Per cent
1	No problem	43	71.6
2	Children unemployed	3	5
3	Husband is drunker	5	8.3
4	Road problem	2	3.3
5	Husband is sick	1	1.7
6	No male earner	1	1.7
7	House problem	3	5
8	No profit	1	1.7
9	Extra marital relations	1	1.7
	Total	60	100

These women face problems not only at the place of work but also in the house. The unemployed children, alcoholic and/or sick husband, extra marital relations of the husband, physical harassment by the husband, lack of proper housing facilities, etc., are the important problems faced by the fisherwomen. Many of them shoulder the entire responsibility of running the family. These women feel that they are too much pre–occupied with their work and cannot give much attention to their family. Like other Indian women, they have to attend to their household chores along with their fish marketing activities.

VARIOUS ACTIVITIES OF THE FISHERWOMEN

Whether selling fish in the markets or not, women are responsible for domestic chores and child rearing, although some men help in some household activities. Fisherwomen in the joint family share household duties. Elder daughters or younger sisters look after children, clean and cook. However, fisherwomen as they grow older, play important role in the economic activities such as processing, cleaning, salting, drying and selling of fish both wholesale and retail in the markets. In the olden times, i.e., before the introduction of nylon nets, women were often responsible for net-making and repair. The women engaged in net making have been marginalized with the advent of mechanization during the eighties and are now forced to work as wage earners in other areas (Nair: 1998). Because, the demand for hand made nets is falling rapidly with the increased mechanization. With the advent of trawling and increased catches since 1975, the entrepreneurial Goan fisherwomen have ventured into salting and drying activities. As a result of growing economic power and accumulation of wealth, their individual sense of pride and identity is enhanced. This has significantly contributed to the enhancement of their family status. Their economic success can be seen in their improved living conditions such as better diet, health care, housing conditions and better educational opportunities for their children.

OCCUPATIONAL PROBLEMS FACED BY FISHERWOMEN

Some of the major problems faced by the fisherwomen in their occupations are discussed below:

(1) Lack of raw- material

Gone are the days when the fisherwomen could take the fish caught by the male members of her family and sell it in the market. With the kind of present sharing systems, the auctioning system, the shortage of fish and the migration of fishermen, fisherwomen nowadays are forced to compete with the traders and merchants at the beach to buy a basket of fish for retailing and most often they are left with no fish. Even then, they are marginalized by traders and agents of export merchants who do not even belong to the fishermen community.

(2) Lack of basic facilities

The poor fisherwomen who venture out of their houses in the wee hours of the morning have no decent place even to relieve themselves during the course of the day. These fisherwomen are seen washing their hands and utensils under roadside taps or in filthy wayside water after selling the fish.

(3) Lack of adequate transportation facilities

Though the fisherwomen have to travel for miles together to purchase and sell fish, little thought has been given to provide alternate transportation facilities as they are very often not allowed to use the public transport system. This has resulted in acute wastage of time and energy of these fisherwomen, which if otherwise harnessed, could be more productive.

(4) Shortage of working capital

Though the commercial banks have now come forward to assist the fisheries sector, still the amount that can be mobilized through institutional credit is meager in comparison to the demand. This has pressurized the fisherwomen to go to the private moneylenders for meeting their short-term working capital requirements at exorbitant rates of interest.

(5) Lack of storage facilities

Fish, being a perishable commodity, has to be disposed of at the earliest. This urgency to sell the fish as fresh as possible to get the maximum return coupled with the lack of adequate transportation facilities often forces the fisherwomen to give away the fish at throwaway prices at the end of the day. The lack of even simple and small storage facilities is acutely felt at such times.

(6) Leakages in Income

(1) Interest on loans taken from non-formal sources

The credit taken from non-formal sources attracts exorbitant rates of interest, sometimes-even upto 30 per cent per day. This leads to a severe drain on the income of the women and thereby affects the income of the household as a whole.

(2) Transportation Charges

Since the fisherwomen vendors are not allowed to travel with fish in the State Transport buses, they have to seek other forms of transport, the charges of which are very high. But the necessity to sell fish at the earliest in order to get maximum returns and the lack of adequate cheap transport facilities forces the fisherwomen to resort to these transportation facilities. A huge portion of their income goes to meet this requirement.

(3) Drinking/ Gambling by men

The surplus generated by the female members of the household after a day's hard work, is very often consumed by the male members for their own entertainment.

No study has also ever tried to quantify the labour of these fisherwomen. This is the crux of the problem whereby the women themselves have such a low esteem of their work and they continue the drudgery without even noticing that they are being exploited.

SUGGESTED MEASURES FOR IMPROVING SOCIO-ECONOMIC STATUS OF FISHERWOMEN

Chaudhary (2002) has made following suggestions for improving socio-economic status of fisherwomen:

(1) To identify infrastructure with reference to the role of women in fishery and related activities and to provide such support to them by way of cold storage, processing space, transportation network, and sales outlets and peeling sheds etc. under cooperative umbrella.

(2) In every field there is constant change and upgradation. To suit this in fisheries sector, proper training is needed for women workforce. Necessary training centres may be set up for women in different levels to provide skills through appropriate professional training on different sectors of fisheries with inclusion of capture and culture fisheries. Facilities should be provided in the universities and other research and development institutions to train small groups of fisherwomen in the business of fishing.

(3) During off-season alternative jobs/ vocations to be provided. Schemes like saving- cum- incentives may be strengthened to provide alternative livelihood.

(4) Financial weaknesses are basic drawbacks in fisheries, women are facing the same. Micro level loan facilities are essential to be disbursed through nationalized banks and NCDC sectors.

(5) Socio-economic census for fishermen will be an effective tool for proper assessment of numbers and exact share of women in the sector. Necessary data bank may be created to formulate their development programme.

(6) While making policies for women involvement, NGOs can play a mediator role between government and private sector.

(7) Providing facilities like potable water, roofing, raised platforms and toilets to women in fish markets and landing centres. A Welfare Board exclusively meant for fisherwomen be introduced to look after their insurance, pension benefits, medical facilities, minimum wages, financial aids etc.

(8) Education to fisherwomen is must and should be encouraged.

CHAPTER V

ECONOMICS AND MARKETING OF FISHERIES

The marine fisheries all over the world are characterized by a dualism in the form of co-existence of small scale or artisan fisheries side by side with large scale or industrial fisheries. The dualism is not confined to the scale of operation but extends to the type of technology used, the degree of capital intensity, employment generation and labour-intensive using little capital and hardly any modern technology (Panayotoy: 1982). In this context, it is essential to know the economics of fishing of both the artisan and the mechanized fishermen.

FISHING EFFORTS

A fundamental step in the economic analysis of fisheries is to develop a relationship between the activities of a fishing fleet and the catch it takes. Fishing fleets typically consist of various types of vessels, equipped in different ways and employing crews with different skills. This is partly the result of technological progress making old techniques obsolete in the long term. But the older fleets can still be profitably used before they are worn out. The diversity of fishing fleets is the result of different judgments or preferences among individual decision makers. The type of fishing boat which appears most profitable to one particular fisherman or fishing firm might not be seen that way by all other fishermen or fishing firms (Hannesson: 1993).

There is a lot of variation in the fishing efforts put in by the fishermen in the study area. The traditional fishermen themselves row their crafts with the help of one or two labourers or some family members. How far and fast he can go in the sea depends upon their physical effort, stamina and patience. The motorized fishermen have the motors fitted to their crafts and hence, they can easily go far off in the sea. The trawling boat owners employ two tandels, five khalasis, and a cook to carry out the fish catching task. Some times, they stay for even eight days at sea. While the purseine boat owners employ two tandels (drivers), two aryamen (persons who cast nets), 18 khalasis (sailors) and a cook on their boats. The basic difference between a trawling boat and a purseine boat is that of the nets and not the craft. The number of labourers employed on the trawling boat and the purseine boat also varies as mentioned earlier.

THE CONCEPTS OF REVENUE, COST AND PROFIT

Revenue is the price of a product multiplied by the quantity sold. Symbolically, $R = (P \times Q)$ where R is revenue, P is price and Q is quantity sold. A single seller who contributes only a small fraction of the total supply of fish in a given market cannot influence the price to any appreciable extent by varying the amount sold. Each agent in a market gets fish from many 'small' suppliers. The price is an outcome of the interaction among many agents, on which each one has little or no influence. From the aggregate point of view this is not the same. The price of fish may vary greatly with the amount of fish bought and sold within the relevant period in a day, a week, or a month. This is to be taken into account while considering the aggregate revenue of the fishermen.

The present study found that the fishermen who have more catch and who have taken loans from the agents have to accept the price quoted by the agents. When the catch is more of a particular variety, lower price is quoted by the agents and vice-versa. Whereas the traditional fishermen who are not caught in the debt trap have full freedom to quote their price. This is true when mechanized fishermen do not go for fishing particularly during the ban period. But since their catch is small in quantity, it hardly influences the total supply and the price level at a particular point of time. It is, therefore, can be generalized that the revenue earned varies by the season, fishermen to fishermen and landing centre to landing centre. A place that is closer to the market area will fetch a higher price and higher revenue for the fishermen and vice-versa.

The analysis of cost incurred by the fishermen is somewhat a difficult task. One basic distinction to be made is between long-term and short-term costs. Short-term costs, or variable costs, are such cost items that can be varied at short notice. Fuel cost is probably the best example of this; whether or not to buy fuel, and how much, typically is a decision that can be taken at short notice, daily or weekly, depending on the reliability of supplies. Labour costs may be of this kind as well; it may be possible to hire labour even on a daily basis, but typically one has to provide employment for a longer period (paying Labour a share of the catch value makes labour cost a short-term cost). The cost of a boat on the other hand is a typical long-term cost, or fixed cost. The boat has to be paid for when acquired, typically with a loan which must be paid back in instalments and on which one must pay interest, regardless of whether the boat is being used or not.

Profit is the difference between revenue and cost the fishermen have to incur. Symbolically, (π = R - P) where, R is revenue, P is price and Đ is profit. Depending on whether one takes a long or a short-term view we can speak about long-term and short-term profit. Looking only at the short-term one would subtract only short-term costs from the revenue, so short-term profit would include the contribution to paying for durable equipment like boats. If the short-term profit exceeds this payment, there would be some long-term profit as well. It could be profitable in the short-term to operate a fishing boat even if the long-term profit is negative. As long as there is some short-term profit making some contribution to paying the fixed cost that must be paid anyway, it obviously makes sense to operate the boat rather than laying it up. On the other hand, it would not be profitable to renew the equipment under such circumstances, as its cost could not be recovered.

Majority of the boat owners in Goa revealed that there are no prospects for long-term profit. Therefore, they feel that they will have to keep the boats idle because the fish resource is depleting over the years.

Most of the fishermen do not maintain any account about the cost. Hence, it becomes difficult to make an analysis of costs of the fishermen. The fishermen have to incur heavy costs on the crafts, nets and other accessories. The table given below shows the cost of crafts and nets used by different types of fishermen.

Table- 5.1 Cost of Crafts and Nets used by the Fishermen

Sr. N.	Type of Craft	Cost of Craft	Cost of Motor	Cost of Engine	Cost of Net	Total Cost
I	**Traditional Crafts**					
1.	Wooden (36 ft)	70,000	_	_	30,000	1,00,000
2.	Wooden (25 ft)	40,000	_	_	30,000	70,000
3.	Fiber (36 ft)	1,30,000	_	_	30,000	1,60,000
4.	Fiber (25 ft)	80,000	_	_	30,000	1,10,000
II	**Motorized Crafts**					
1.	Wooden (36 ft)	70,000	75,300 (9.9 hp)	_	24,000	1,69,300
2.	Wooden (25 ft)	40,000	58,000 (8 hp)	_	12,000	1,10,000
3.	Fiber (36 ft)	1,30,000	75,300 (9.9 hp)	_	24,000	2,29,300
4.	Fiber (25 ft)	80,0000	58,000 (8 hp)	_	12,000	1,50,000
III	**Mechanised Crafts**					
1.	Purseine Boat (60ft)	18,00,000	_	4,00,000	7,00,000	29,00,000
2.	Trawler (60 ft)	18,00,000	_	4,00,000	1,00,000	23,00,000
3.	Trawler (40 ft)	8,00,000	_	3,50,000	1,00,000	12,50,000
4.	Trawler (32 ft)	6,00,000	_	2,00,000	1,00,000	9,00,000

Source: Field Survey and Directorate of Fisheries.

The data about costs of crafts and nets reveal that the fishermen from the mechanised sector have to incur heavy costs on the crafts, nets and engine. The purseine boats incur around 29 lakhs followed by the trawlers, which are also known as fishing boats. The difference between these two boats is the difference of nets. Purse seine net is costlier and is different from the trawling net. Vessels using purse seine are equipped with pursing gallows and pursing winches for hauling the purse lines, which close the net after setting. The fish caught by the purse seine boats is generally more than other trawlers.

The data presented in table 5.2 shows craft wise total fixed costs. The total fixed cost is taken as the sum total of depreciation of engine boats and nets which is worked at 25 per cent of the total value, interest on capital, insurance charges and maintenance of boat, net repairs, replacements, etc. The total fixed cost of purse seine boats is highest compared to all the fishing crafts and that of traditional crafts is lowest

Table – 5.2 Total Fixed Cost of Different Crafts Per Annum (Rs.)

S. N.	Type of Boat	Depreciation P.A. of Engine, Boats Nets etc.	Interest on Capital	Insurance Charges	Maintenance of Boat, Net Repairs & Replacement	Total Fixed Cost
1	2	3	4	5	6	7
I	**Mechanised**					
1.	Purse seine boat (60 ft)	7,25,000	61,625	50,000	2,40,000	10,76,625
2.	Trawler (60 ft)	5,75,000	48,875	58,000	1,40,000	8,21,875
3.	Trawler (40 ft)	3,12,500	26,562	40,000	1,20,000	4,99,062
4.	Trawler (32 ft)	2,25,000	19,125	30,000	1,10,000	3,84,125
	Motorised					
5.	Wooden (36 ft)	42,325	3,598	3,500	12,000	61,423
6.	Wooden (25 ft)	27,500	2,338	2,000	12,000	43,838
7.	Fibre (36 ft)	57,325	4,872	6,500	12,000	80,697
8.	Fibre (25 ft)	37,500	3,187	4,000	12,000	56,687

1	2	3	4	5	6	7
	Traditional					
9.	Wooden (36 ft)	25,000	2,125	3,500	10,000	40,625
10.	Wooden (25 ft)	17,500	1,487	2,000	10,000	30,987
11.	Fibre (36 ft)	40,000	3,400	6,500	10,000	59,900
12.	Fibre (25 ft)	27,500	2,338	4,000	10,000	43,838

Source: Directorate of Fisheries.

Graph 5.1: Total fixed cost of different crafts (P. A. in Rs.)

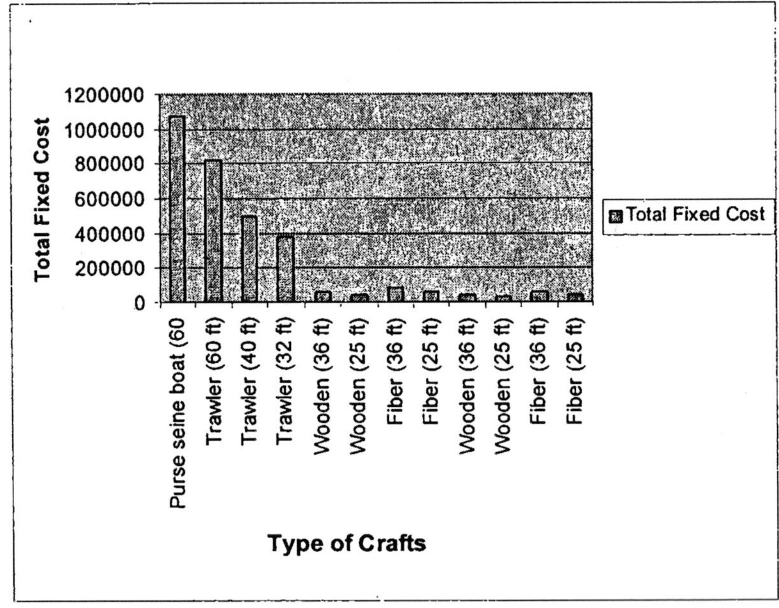

In addition to fixed costs, the fishermen have to also incur a huge amount of variable costs which is shown in table 5.3

An important component of cost is wages paid to the labour either in cash or in kind. Most of the respondents said that they have to get labour from outside, as they cannot get labour in Goa. Average salaries paid to different types of workers are as follows:

Table – 5.3 – Total Variable Costs of Different Crafts per annum (Rs.)

Sr. No.	Type of Craft	Variable Costs						Total
		Fuel Diesel/or	Ice	Labour Kerosene	Food	Transport	Marketing Changes	Variable Cost
I	**Mechanized**							
1	Purse seine boat	1,80,000	4,50,000	9,00,000	9,0000	1,35,000	45,000	25,65,000
2	Trawler 60 ft	1,80,000	7,20,000	4,50,000	36000	45,000	4000	14,31,000
3	Trawler 40 ft	1,40,000	54,000	35,000	32,000	40,000	4000	3,05,000
4	Trawler 32 ft	1,20,000	40,500	30,000	30,000	30,000	3000	2,50,500
II	**Motorized**							
5	Wooden 36 ft	18,000	22,500	90,000	27,000	9,000	8000	1,66,500
6	Wooden 25 ft	16,000	20,000	60,000	18,000	5000	4000	1,19,000
7	Fibre 36 ft	18,000	2,3500	90,000	27,000	9,000	8000	1,75,500
8	Fibre 25 ft	16,000	20,000	60,000	18,000	5,000	4000	12,3000
III	**Traditional**							
9	Wooden (36 ft)	–	–	80,000	–	6,000	–	86,000
10	Wooden (25 ft)	–	–	60,000	–	5,000	–	65,000
11	Fibre (36 ft)	–	–	80,000	–	6,000	–	86,000
12	Wooden (25 ft)	–	–	60,000	–	5,000	–	65,000

Graph 5.2: Total Variable Costs of Mechanized Fishermen Crafts P.A. (in Rs.)

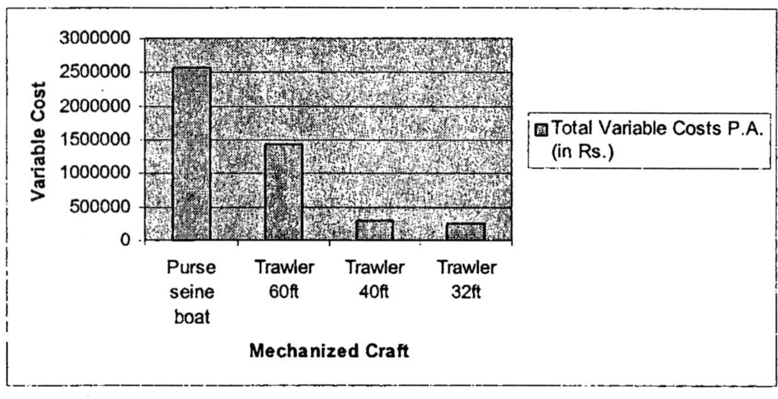

Graph 5.3: Total Variable Costs of Motorized Crafts P.A. (in Rs.)

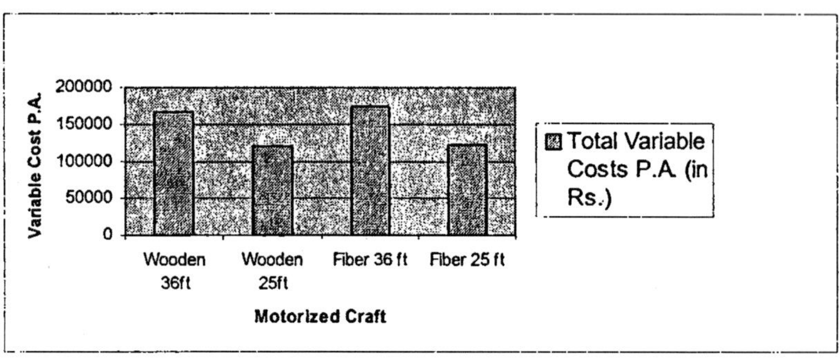

Graph 5.4: Total Variable Costs of Traditional Fishermen Crafts P.A. (in Rs.)

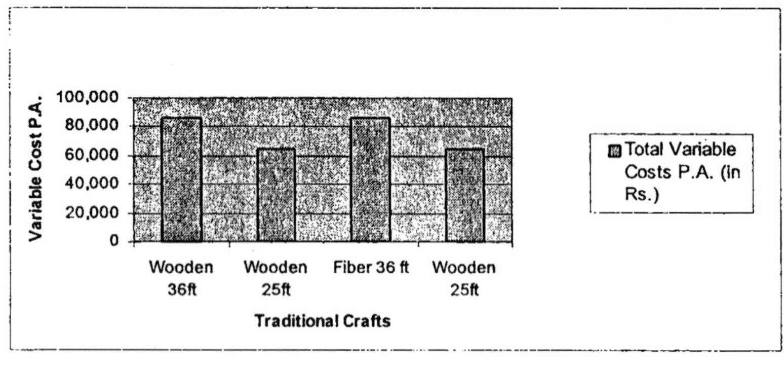

The total variable cost of a purse seine boat is highest, i.e., Rs. 25,65,000 p.a. and that of a traditional craft is the lowest. A traditional fisherman requires a craft, net and a few labourers. The labours employed in the traditional and motorised sector are usually paid in kind i.e. fish given on the basis of sharing system. The whole catch is divided into two equal parts. One part is kept for the boat and nets and the other half is divided equally among the labourers. If the owner also goes for fishing, he also gets a share as other labourers get it.

An attempt was made to quantify the cost of labour in this study and the values were approximated for a comparative analysis of the costs of all the three sectors. There is a lot of variation in the cost structure of the fishermen from different sectors. This variation is mainly due to the difference of crafts, nets, engine, labour, use of fuel and ice.

Table 5.4 gives data about total cost, revenue and profit earned by the fishermen on different types of crafts.

Table – 5.4 - Total cost, Revenue and Profit of different types of crafts (Rs.) per Annum.

S. No.	Type of Boat	Total F.C.	Total V.C.	Total Cost	Gross Revenue	Gross Profit
1	2	3	4	5	6	7
I	**Mechanized**					
1	Purse Seine (60 ft)	10,76,625	25,65,000	36,41,625	54,00,000	1,75,8375
2	Trawler (60 ft)	8,21,875	14,31,000	22,52,875	36,00,000	1357125
3	Trawler (40 ft)	4,99,062	3,05,000	804062	18,00,000	995938
4	Trawler (32 ft)	384125	2,50,500	634625	12,20,000	585375
II	**Motorized**					
5	Wooden (36 ft)	61423	1,66,500	227923	4,00,000	172077
6	Wooden (25 ft)	43838	1,19,000	1,62,838	2,60,000	97162

1	2	3	4	5	6	7
7	Fibre (36 ft)	80,697	1,75,500	256197	4,00,000	1,43,803
8	Fibre (25 ft)	56687	1,23,000	179687	2,80,000	100313
III	**Traditional**					
9	Wooden (36 ft)	40625	86,000	126625	1,80,000	53375
10	Wooden (25 ft)	30987	65,000	95,987	120,000	24,013
11	Fibre (36 ft)	59900	86,000	1,45,900	200,000	54100
12	Fibre (25 ft)	43838	65,000	108838	1,40,000	31162

Graph 5.5: Total cost, Revenue and Profit of Mechanized crafts (Rs.) P. A.

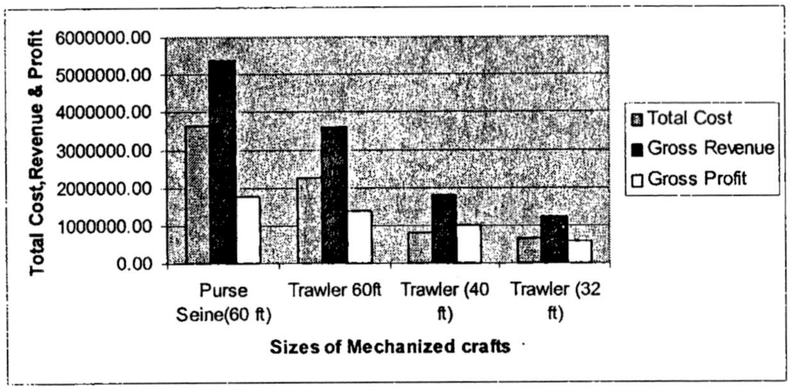

Graph 5.6: Total cost, Revenue and Profit of Motorized crafts (Rs.) P. A.

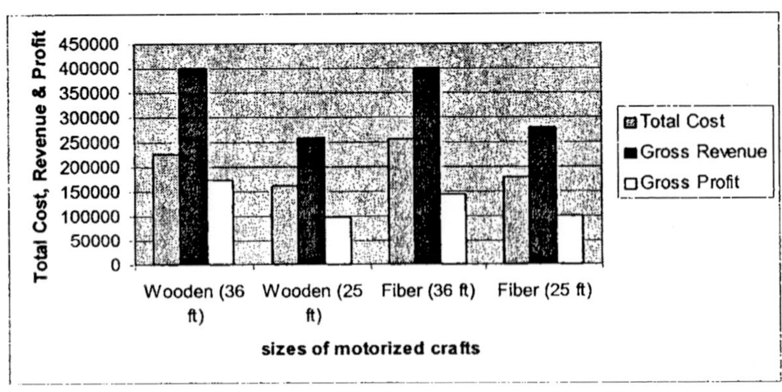

Graph 5.7: Total cost, Revenue and Profit of Traditional crafts (Rs.) P. A.

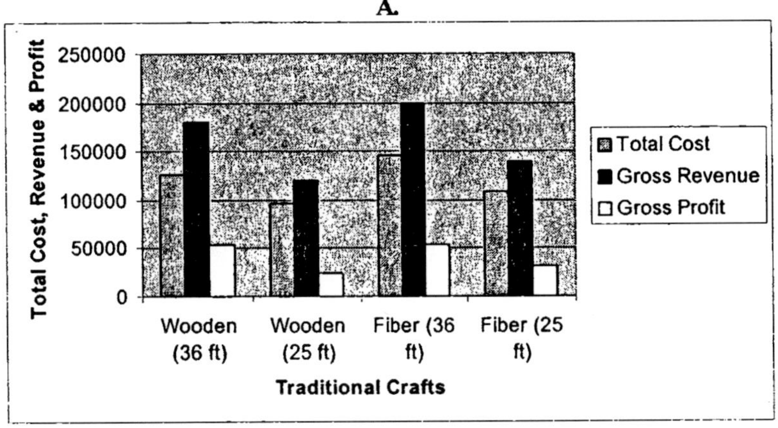

The total revenue and profit varies with the type of crafts. The type of crafts owned by the fishermen depends mainly on their ability to pay for the crafts and enterprising skills. Though, the total cost of traditional fishermen is low, their gross revenue and consequently profit is also low. The increased numbers of crafts, trawlers catching fish close to the shore and thus encroaching in the area which is meant for the traditional fishermen are some of the factors responsible for the deteriorating conditions of the traditional fisher folk. Even the trawler owners complain about falling revenue and profit due to the over utilization of fish resource and excess efforts to catch too few fish.

The cost of fishing varies from craft to craft and from fisherman to fisherman depending upon the efficiency of the entrepreneur and the labourers. The role of Aryaman and Tandel is more important on the mechanized crafts. The experienced and skilled Aryamen and Tandels can make right decisions about. Wage is an important component of the cost of fishing.

It is important to know the mode of fishing operation to understand the wage component in the total cost.

Table 5.5 depicts the mode of fishing operation of the different categories of fishermen in Goa.

Table – 5.5 - Mode of fishing operation

Sr. No.	Mode of operation	Traditional No.	Traditional %	Motorized No.	Motorized %	Mechanized No.	Mechanized %	Total No.	Total %
1	Personally operated	12	12.9	-	-	-	-	12	3
2	Personally with family members	5	5.4	14	9.8	-	-	19	4.75
3	Personally with labourers	31	33.3	83	58	25	12.2	139	34.75
4	Only labourers	44	47.3	46	31.5	105	64	195	48.75
5	Lease out	1	1.1	1	0.7	36	22	38	9.5
	Total	93	100	143	100	164	100	400	100

The profit made by the fishermen also varies with the mode of operation. If the owner of the craft and his family members go for fishing along with labourers, his revenue and profit earned will be more. If the craft is operated by labourers only, the share of the labourers in total catch is more and consequently, the owner's share is less.

LABOUR IN FISHING

The fishing industry requires a large number of labourers. Most of the labourers employed in Goa's fishing sector are the migrants. The reason of employing migrant labourers is that the local labourers do not come forward to take up fishing work. Besides, they are not regular. The labour brought from outside remain for at least one season. Most of the labourers in the study area come from Orissa, Karnataka, Maharashtra, Andhra Pradesh and Tamil Nadu. They migrated to Goa as they were unemployed at the place of origin. We were surprised to know that some of them had passed SSC and three of them had completed their graduation.

Aryaman is paid the highest compared to other labourers on the trawler, because he is the important person who casts net at the right place in the sea. Based on the movement of the water, he can make out the type of fish available in a particular area. With his expertise, he can see whether it is a high value fish or low value fish and accordingly, he decides whether to cast the net or not. Tandel is also another important person on the trawler who decides whether to go far off or not based on the currents of seawater. The labourers are usually given a rise in salary around Rs. 100 per year.

Table 5.6 shows the average monthly salaries paid to workers working on a purse seine boat.

Table 5.6 –Average Monthly Salary paid to Workers on a Purse seine Boat (Rs.)

Sr. No.	Designation of workers	Rank	No. of workers	Wage (Rs.)	Tips/ Allowance	Total Amount (Rs.)
1	Tandel	1	1	10,000	500	10,500
2	Tandel	2	1	8,000	500	8,500
3	Maskman	1	1	7,000	500	7,500
4	Maskman	2	1	6,000	500	6,500
5	Aryaman	1	1	12,000	2000	14,000
6	Aryaman	2	1	10,000	1000	11,000
7	Khalasis	-	18 @ Rs. 3000 per khalasi	54,000	2000 (to be shared)	56,000
8	Cook	-	1	3000	100	3100
	Total		25	110,000	7,100	1,17,100

The table 5.6 gives average salary paid by a purse seine boat owner to his workers. The wage structure depicted in the table given above is of a boat, which is operating for nearly 8 to 10 years with the same workers. Generally, the workers get an increment in the salary by at least Rs. 100 per year. The Aryaman on a new boat gets around Rs. 6000 per month. Based on his fish detecting skill and experience, he gets increments in the salary.

The number of labourers working on a trawler is less than that of the number of labourers on a purse seine boat.

Table 5.7 shows the number of labourers working on a trawler.

Table 5.7 Number of labourers working on a trawler

Sr. No.	Designation of Workers	Rank	No. of Workers
1	Tandel	1	1
2	Tandel	2	1
3	Khalasis	-	5
4	Cook	-	1
	Total	-	8

There is no fixed wage to the labourers on a trawler. Generally, Tandel of Rank 1 distributes 30 per cent of the catch value to all the workers after keeping his share. He acts as the leader of all the labourers. The remaining 70 per cent of the catch value is for the boat owner towards fixed costs and variable costs.

PROBLEMS FACED WITH REGARDS TO LABOURERS

The fishermen in Goa face a number of problems with regards to labourers.

Table 5.8 gives an account of various problems faced by the respondents

Table – 5.8 - Problems faced with regards to labourers

Sr. No.	Problems	Traditional		Motorized		Mechanized		Total	
		No.	%	No.	%	No.	%	No.	%
1	Lack of skill	20	21.5	5	3.5	6	3.7	31	7.7
2	Absenteeism	12	12.9	31	21.7	33	20.1	76	19
3	Labour turnover	11	11.8	25	17.5	28	17.1	64	16
4	Demand high wages	19	20.4	28	19.5	22	13.4	69	17.3
5	Inefficiency	10	10.8	-	-	10	6.1	20	5
6	Running with advances	-	-	-	-	20	12.2	20	5
7	Alcoholism	-	-	24	16.8	26	15.8	50	12.5
8	No specific problem	21	22.6	30	21	19	11.6	70	17.5
	Total	93	100	143	100	164	100	400	100

The more prominent problems faced with regard to labourers across all the 3 sectors are absenteeism, poor labour turnover and demand for higher wages. Many of them in all the 3 sectors, however, also said that they had no specific problems.

FINANCING OF FISHERIES

The fishermen raise finance from two sources namely, internal finance and external finance.

Table 5.9 shows the pattern of internal finance of respondents.

Table - 5.9 - Pattern of Internal Finance of some respondents.

Sources	Sectors						Grand Total
	Traditional		Motorised		Mechanised		
	Total Amount	No.of fisher-men	Total Amount	No.of fisher-men	Total Amount	No. of fisher-men	
Own Capital	461400	24	329000	7	1790000	22	2580400
Family Property	30000	2	45000	2	380000	4	455000
Marriage gift & others	20000	7	—	—	120000	5	140000
Total	511400	33	374000	9	2290000	31	3175400

There are three sources of internal finance available to the fishermen – own capital, family property and finance from marriage gift and others. Majority of the fishermen in all the three sectors: traditional, motorized and mechanized have a major part of their internal finance coming from their own capital. A very small part of their capital comes from the other two sources. Also, a very small number of the total number of fishermen depends on these two sources. Fishermen in the motorized sector have their internal finance coming from only the first two sources. The fishermen borrow from different sources for meeting their costs.

Table 5.10 reveals the pattern of external finance of some fishermen. Very few of the fishermen from all the three categories are indebted to the external agencies.

Table - 5.10 - Pattern of External Finance

Sources	Sectors						Grand Total
	Traditional		Motorised		Mechanised		
	Total Amount	No.of fisher-men	Total Amount	No.of fisher-men	Total Amount	No. of fisher-men	
Comm. Bank	1685500	25	5144000	63	42815000	75	49644500
Co-op. Bank	-	-	150000	3	995000	6	1145000
Friends	231000	4	40000	2	70000	1	341000
Relatives	-	-	-	-	119000	4	119000
Not borrowed	-	64		75		78	-
Total	1916500	93	5334000	143	43999000	164	51249500

There are four sources of external finance available to the fishermen –Commercial Banks, Co-operative Banks, Friends and relatives. Majority of the fishermen in all the three sectors depend on Commercial Banks for a major part of their external finance. Few of the traditional fishermen also depend on friends for a small part of the finance. While those in motorized sector depend on Co-operative societies as well as friends, those in mechanized sector depend on all the other three sources for a part of their external finance requirements. The fishermen have to keep some items as security to obtain loans from the financial institutions.

Table 5.11 shows the items of security to get loans. The above data shows that the traditional fishermen have on an average indebtedness of only about Rs. 773, while in case of motorized fishermen an average indebtedness is Rs. 30910. The extent of indebtedness is very high for fishermen in the mechanized sector, i.e., Rs. 152723.

Table- 5.11 – Items of security to get loans

Sr. No.	Items of security	Traditional		Motorized		Mechanized		Total	
		No.	%	No.	%	No.	%	No.	%
1	Craft/ Canoe	5	5.4	16	11.2	30	18.3	51	12.7
2	House	2	2.2	3	2.1	9	5.5	14	3.5
3	LIC Policy/ FD	8	8.6	12	8.4	19	11.6	39	9.6
4	Nets	8	8.6	24	16.8	20	12.2	52	13
5	Gold	6	6.5	13	9.1	8	4.9	27	7
6	Not borrowed	64	68.7	75	52.4	78	47.5	217	54.2
	Total	93	100	143	100	164	100	400	100

The fishermen have taken loans on security of LIC Policy/ Fixed Deposit receipts and also gold to certain extent. Some fishermen in motorized and mechanized sectors have taken loans on security of the craft/ canoe.

In recent years the fish resources have been depleting. The fishermen, therefore, face a number of problems in repayment of loans as shown in table 5.12.

Table- 5.12 - Problems faced in repayment of loan

Sr. No.	Problems	Traditional		Motorized		Mechanized		Total	
		No.	%	No.	%	No.	%	No.	%
1	Low catch	21	23	68	48	60	37.6	149	37.3
2	Low Price	18	19	-	-	-	-	18	4.5
3	Ban period	-	-	-	-	14	8	14	3.5
4	Non-operational Boat	-	-	-	-	4	2.4	4	1
5	High diesel prices	-	-	-	-	8	5	8	2
6	Not borrowed	54	58	75	52	78	47	207	51.7
	Total	93	100	143	100	164	100	400	100

The fishermen who have borrowed loans face problems in repayment of loans because of low catch of fish. Because the catch is less, their profit is less and they cannot repay the loans.

Fishing is a risky occupation. Weather conditions and sea waves keep fluctuating. The fishermen who go for fishing may be caught in the disturbed sea and some times, may not return home. They require clear information about weather conditions before making a decision to go for fishing or not on a particular day. Table 5.13 shows the sources of information to fishermen about weather conditions.

Table - 5.13 - Sources of information to fishermen about weather conditions

Sr.	Source of	Traditional		Motorized		Mechanized		Total	
No.	Information	No.	%	No.	%	No.	%	No.	%
1	Media	29	31.1	11	8	30	18.2	70	17.5
2	Exp./own ex.	60	64.5	80	55.9	80	49	220	55
3	Satellite	1	1.1	6	4	2	1.2	9	2.3
4	Union	-	-	2	1.4	36	22	38	9.5
5	Fishing federation	-	-	-	-	1	0.6	1	0.2
6	Newspaper	1	1.1	2	1.4	4	2.4	7	1.7
7	Wireless	1	1.1	-	-	3	1.8	4	1
8	Directorate of Fisheries	-	-	-	-	1	0.6	1	0.2
9	Fishermen societies	-	-	40	27.9	1	0.6	41	10.3
10	Local fishermen	1	1.1	2	1.4	6	3.6	9	2.3
	Total	93	100	143	100	164	100	400	100

More than 50 per cent of the respondents do not depend on other sources to know about weather conditions. They come to know about the sea and weather conditions as they have a long past experience. This trend is seen in all the sectors.

Besides the above source, most fishermen in traditional sector depend on media, in motorized sector on fisheries survey and in mechanized sector on their union as well as media for this information.

Fev thers, in all the 3 sectors, depend on the other sources like satellite, newspaper, wireless, etc.

The fishermen make some preparation before sailing out for fishing, Table 5.14 shows what preparations are made by the respondents before fishing.

Table - 5.14- Rituals and Practices followed by Fishermen before fishing

Sr.	Preparations	Traditional		Motorized		Mechanized		Total	
No.		No.	%	No.	%	No.	%	No.	%
1	Prayers	41	44	22	15	51	31.1	114	28.5
2	Repairs	37	40	90	63	86	52.4	213	53.3
3	Pooja	15	16	31	22	27	16.5	73	18.2
	Total	93	100	143	100	164	100	400	100

Majority of the fishermen in all the three sectors do repair work before fishing. Many of the others in traditional sector as well as mechanized pray God for safety and good catch, while those in motorized sector perform pooja before going for fishing.

MARKETING OF FISH

The growth of fish production and development of fishery sector is dependent on the efficient fish marketing system. The post-harvest operations of fish provide more employment to labour than the production sector. Improved methods of handling and storage of fish in recent years has led to rapid changes in the distribution process and fish marketing system (Sathiadhas and Kanagam: 2000). The fresh fish that were inaccessible to far off places from landing centres a few years back are now easily accessible due to the vast improvement in handling technology coupled with fast transportation facilities and consequent market penetration. Though, the iced fish faced consumer resistance initially, now it has got almost total consumer acceptance.

However, both in internal and external marketing, a large number of intermediaries are involved before fish reaches the hands of the ultimate consumer. An efficient marketing system of any commodity aims at ensuring the services of middlemen at minimum cost.

Fish marketing may be broadly defined as all those functions involved from the point of catching of fish to the point of final consumption. As the fish, like any other product moves closer and closer to the ultimate consumer, the selling price increases since the margins of the various intermediaries and functionaries are added to it. The pricing efficiency is concerned with improving the operation of buying, selling and other connected aspects of marketing process so that it will remain responsive to consumer direction.

Common problems found in fish distribution and marketing are as follows:

1. Greater uncertainties in fish production and hence in the supply of fish.
2. High perishability of fish.
3. Assembling of fish from too many coastal landing centres.
4. Too many varieties and hence, too many demand patterns.
5. Wide spatial and temporal variations in market arrivals and prices.
6. Disequilibrium of demand and supply.
7. Difficulty in maintaining the quality of fish.
8. Lack of information on fish price and production.

The efficiency of fish marketing will be high if these problems are solved so that both the fishermen and the consumers will be satisfied fully.

FISHERMEN'S SHARE IN CONSUMER'S RUPEE

Fishermen's share in consumer's rupee is the best index to measure the efficiency of fish marketing system (Sathiadhas and Kanagam: 2000). It is a well known fact that the fishermen do not receive legitimate share of the increased price paid by the consumers due to the larger magnitude of marketing margins. Higher the value of marketing margins lower is the efficiency of marketing system (Sathiadas and Panikkar: 1992). The perishable nature of fish, uncertainties in fish landings, assembling of fish food, too many coastal landing centres, too many demand patterns and transportation of fish to different regions and interior areas without affecting the quality are same of the key issues in marine fish marketing (Rao: 1983).

As the fish requires the quickest possible movement from the landing centre to retail markets, a number of transactions are involved before it reaches the ultimate consumer.

Sathiadas and Kanagam (op cit) worked out the percentage distribution of consumer's rupee for different varieties of fish during 1996-97 on all India level. They found that the fishermen share in consumer's rupee ranges from 30 to 68 per cent for different varieties. Marketing costs including transportation ranged 6 to 13 per cent of the consumer's rupee. Wholesalers received 5 to 32 per cent and retailers from 14 to 47 per cent of consumer's rupee for different varieties of marine fish.

Table - 5.15 – Percentage distribution of consumer's rupee for different varieties of marine fish in India (1996-97)

Name of fish	Fishermen	Handling & Transportation	Wholesalers	Retailers
1	2	3	4	5
Seer fish	68	6	12	14
Pomfrets	60	7	9	24
Barracudas	40	9	30	21
Tuna	45	9	28	18
Sharks	43	10	32	15

1	2	3	4	5
Catfish	56	10	10	24
Mackerel	50	9	11	30
Sardines	33	12	23	32
Ribbonfish	48	10	12	30
Rays	47	13	22	28
Whitebaits	40	12	28	20
Lizardfish	35	12	15	38
Goatfish	57	13	16	14
Threadfins	42	9	20	29
Croakers	48	11	14	27
Silver bellies	30	15	8	47
Big-jawed jumper	55	10	9	26
Mullets	41	9	17	33
Half & full beaks	65	9	10	16
Cephalopods	65	10	5	20

Source: Sathiadhas R. and A. Kanagam (2000). Pillai V.N. and N.G. Menon (eds). Marine Fisheries Research and Management Cochin: CMFRI. pp. 869

FISHERMEN'S SHARE IN CONSUMER'S RUPEE IN GOA

In each rupee spent by the consumer, the share of fisherman is around 60 per cent and remaining 40 per cent goes to the middlemen. The percentage distribution of consumer's rupee for different varieties of marine fish in Goa during 2004-05 is depicted in table 5.16.

Table – 5.16– Percentage Distribution of Consumer's rupee in Goa for different varieties of marine fish in Goa during 2004-05.

Name of fish	Fishermen	Handling & Transportation	Wholesalers	Retailers
Seerfish	60	20	10	10
Pomfrets	60	20	10	10
Sharks	50	20	15	15
Prawns	65	10	10	15
Butter fish	50	15	15	20
Tuna	50	15	15	20
Kowal kawal	45	15	15	25
Mackerals	55	15	15	20
Sardines	55	15	10	20

The fishermen's share in consumer's rupee ranges from 50 to 60 per cent for different varieties. The marketing costs including transportation ranged from 15 to 20 per cent of the consumer's rupee. The wholesalers received 10 to 15 per cent and retailers received 10 to 20 per cent of consumer's rupee for different varieties of marine fish in Goa.

MARKETING METHODS USED BY THE RESPONDENTS

It is important to know, whether they sell to retailers, consumers, head loaders, agents, company or they follow auction sale method. It was found that only 31 per cent of the fishermen sell fish directly to the consumers. From the traditional category, 48 per cent and from the motorized category 53 per cent of the fishermen sell directly to consumers either through their wives or through any other family members.

Table - 5.17- Methods used for the sale of catch

Sr. No.	Sale	Traditional		Motorized		Mechanized		Total	
		No.	%	No.	%	No.	%	No.	%
1	Retailer	30	32.2	17	12	41	25	88	22
2	Consumer	45	48.4	76	53	3	1.8	124	31
3	Head Loader	16	17.2	33	23.1	3	1.8	52	13
4	Auction	-	-	8	5.6	4	2.4	12	3
5	Agents	2	2.2	9	6.3	86	52	97	24.3
6	Company	-		-	-	27	17	27	6.7
	Total	93	100	143	100	164	100	400	100

From mechanized sector, 52 per cent of the fishermen sell to the Agents. These agents lend loans to the fishermen whenever required and in turn the fishermen are obliged to sell the catch to them. They also sell to the fish processing companies. Only 27 mechanized fishermen sell the catch to the companies. It was found that the agents and the companies quote the prices and hence, the fisherman does not have much freedom in fixing the price. This means that these middlemen take away a part of the income of the fishermen by providing those advances. If the fishermen sell directly in the town markets or to the hoteliers, they can earn more income. Besides, the consumers also can get fish at a lower price. But due to the involvement of the middlemen, neither the fishermen can earn a higher income nor can the consumers get cheaper fish. That means, the middlemen make a good amount of profit at the cost of both the fishermen and the consumers.

MARKETING MARGINS AND EFFICIENCY

The gross marketing margin refers to the difference between the price paid by the consumer and the price received by the producer. This includes all costs of assembling, grading, packing, transportation, processing and storage, wholesalers' and retailers' margin.

The marketing margin is an indicator of the marketing efficiency. Higher the value of the marketing margin, lower is the efficiency of the system. On the one hand, the producers deserve a legitimate share in the consumer's rupee and on the other hand, the consumers rights have to be safe guarded against excessive prices. These twin objectives can be achieved by ensuring various marketing services at reasonable costs, i.e., restricting margins at a reasonable level.

The fishermen lack the marketing facilities near jetties. They face the problem of transportation. Some mechanized and motorized fishermen own tempos and pickups. But many of them have no option but to sell to the agents or companies or to the retailers.

The gross marketing margin of the middlemen and fishermen can be found out by using the following formulae:

Gross Marketing (G.M.) = Retail price (RP) – Landing centre Price (LP)

Percentage share of middlemen in

$$\text{Consumers rupee} = \frac{RP - LP \times 100}{RP}$$

Percentage share of fisherman in

$$\text{Consumer's rupee} = \frac{LP \times 100}{RP}$$

PRICE BEHAVIOUR

The level of supply, consumer's preference, price of other varieties of fish and general price level of vegetables and meat are some of the factors which influence the price of fish. It is observed that there have been considerable variations in the marine fish not only between seasons but also between different days and even on the same day between morning and evenings. The demand for fresh fish is usually high in the morning because buyers are ready to pay a higher price in the morning and do not wait for the increased supply of fish in the later part of the day. The price also varies between peak and lean seasons.

Table – 5.18 – Quarterly minimum and maximum average landing centre price for selected varieties of fish in Goa (Rs. / kg.) During 2004-05.

Name of fish	Traditional		Motorized		Mechanized		Total	
	Min.	Max.	Min.	Max.	Min.	Max.	Min.	Max.
Seerfish	90	120	70	90	80	120	80	103
Pomfrets	110	130	100	110	110	120	106	120
Sharks	60	70	40	50	50	60	50	60
White Prawns (50 count)	200	250	180	200	220	240	200	230
Butter fish	50	70	30	50	50	70	43	63
Tuna	22	25	15	20	30	40	27	37
Kowal kawal	30	40	20	30	30	40	27	37
Mackerals	25	35	24	34	26	36	25	35
Sardines	6	8	4	6	5	7	5	7

It is clear from the table 5.18 that the average landing centre price varies from season to season and from fish to fish. The maximum and minimum price also varies from area to area.

Table – 5.19– Quarterly minimum and maximum average retail prices for selected varieties of fish in Goa (Rs. / kg.) during 2004-05.

Name of fish	March-May		Mid Aug.-Nov.		Dec.-Feb.		Overall	
	Min.	Max.	Min.	Max.	Min.	Max.	Min.	Max.
1	2	3	4	5	6	7	8	9
Seerfish	100	130	90	100	100	120	97	117
Pomfrets	120	140	110	120	120	130	117	130
Sharks	70	80	50	65	70	80	63	75
White Prawns(50 count)	220	270	190	210	230	250	213	243

1	2	3	4	5	6	7	8	9
Butter Fish	60	80	40	60	60	80	53	73
Tuna	25	32	20	25	35	40	27	32
Kowal kawal	35	45	30	40	40	50	35	45
Mackerals	35	45	30	35	35	40	33	40
Sardines	8	10	6	8	7	10	7	9

The average retail price of all the varieties of fish is higher than the wholesale price which indicates the value added by the retailers. Sometimes, the value added is almost the double of a variety if nobody in the market sells that particular variety. But sometimes, they cannot cover the cost at which they had bought from the wholesalers or from the fishermen. These are the exceptional cases. But, in general, they add value to the price and that is their profit.

PRICE DETERMINATION OF FISH

The price of fish fluctuates more than any other agricultural commodity. The price fluctuations take place due to changes in supply of a particular variety of fish and also due to the variations in prices of other fish varieties in the market (Sathiadas and Narayana Kumar: 1994). The variations of fish prices at all stages of transactions are attributed to the uncertain nature of fish production and perishability.

Prices of fish are determined by the interaction of demand and supply conditions both at producing centres and consumer markets. At the lending centres, the market demand is the aggregate demand from wholesalers and agents which is indicated by the number of trucks, cylinders, retailers and individual purchases arrived at the centre. The short run demand is more or less stable. Often the demand for fish is high either in the morning or in the evening hours. Sometimes, the retailers are prepared to pay high price for fish in the morning without worrying about the increased supply at the later part of the day. The short run supply of fish is highly inelastic and unpredictable.

MARKETING CHANNEL

Marketing Channel implies the path through which the product passes from the producer to the hands of the ultimate consumer. As far as marine fish is concerned, fish travels long distances from coastal areas to the interior parts of the state.

The usual marine fish marketing channels are shown aptly by Sathiadas and Narayankumar(1994) which are applicable in all the regions. They are:

1. Fisherman-Auctioneer-Agents of freezing plants-Exporters.
2. Fisherman-Auctioneer-Processor (Dry Fish) – Wholesaler-Retailer-Consumer.
3. Fisherman-Auctioneer-Wholesaler (Primary Market) – Wholesaler (Retail Market) – Retailers-Consumers.
4. Fisherman-Auctioneer-Commission Agents-Wholesaler-Retailer-Consumers.
5. Fisherman-Auctioneer-Retailer-Consumer.
6. Fisherman-Auctioneer-Consumer.

The major portion of the internal fish marketing takes place through 3-6 channels. The auctioneers of the primary market and commission agents of the secondary market are also involved in the process without taking possession of the fish.

Figure 5.1: Flow Chart showing the fish marketing channels

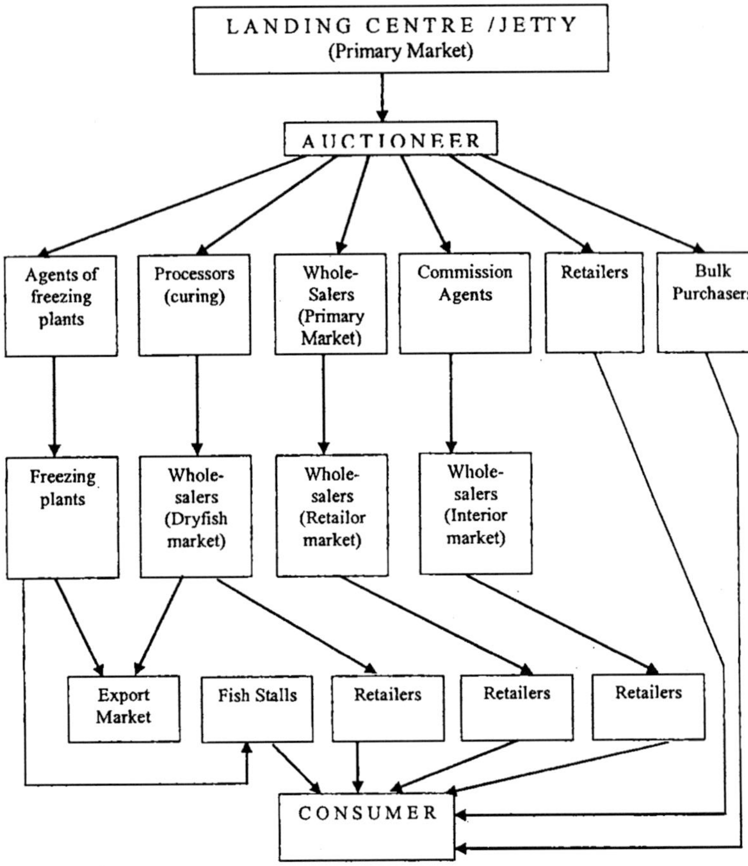

LANDING CENTRE /JETTY
(Primary Market)

AUCTIONEER

Agents of freezing plants

Processors (curing)

Whole-Salers (Primary Market)

Commission Agents

Retailers

Bulk Purchasers

Freezing plants

Whole-salers (Dryfish market)

Whole-salers (Retailor market)

Whole-salers (Interior market)

Export Market

Fish Stalls

Retailers

Retailers

Retailers

CONSUMER

Source: Sathiadas R. and R. Narayana Kumar (1994). "Price policy and fish marketing in India", Biology Education / Oct. – Dec. pp. 231.

In Goa, fish marketing has not developed fully on modern lines. There is a gradual transformation from traditional to modern method of marketing with the advent of improved transport, processing and storage facilities. At micro level, there is large number of small merchants. At macro level few organizations undertake the whole distribution of fish.

At landing centres, fishes are disposed by auctioning. This provides maximum competition among buyers and enables quick disposal. Usually at landing centres, fish is not sold in weight because of the practical difficulties involved in the handling of such a highly perishable commodity. Hence, the sales are carried out by measures of heaps or lots of different sizes. However, for exportable varieties like prawns, the price per kg of fish is fixed by auction and weighed before delivery. Generally, traditional auctioneer or middlemen on commission basis who take up the responsibility of realizing the sale proceeds from the traders are involved in auctioning. The auctioneers at the landing centre take 5-10 per cent of fish auctioned by them as commission. Since many of the auctioneers advance loans to the fishermen, they take a portion of share towards the interest for the loan given.

CHAPTER VI

IMPACT OF MECHANIZATION ON FISHING

The mechanization of the Indian fishing industry began with the inception of the Indo-Norwegian Project in Kerala in 1953. The objective was to improve the standard of living of the population by increasing the fish catch and thereby, the protein available to the people for consumption. It also aimed at improving the economic conditions. Mechanization was to increase not only the catch but also productivity by the use of modern equipment and technology. The hidden motive seems to have been the desire of the West to create a market for its fishing equipment and technology (Alvares: 2002).

MECHANIZATION IN MARINE FISHERIES

There have been variations in the type of mechanization. Innovations in the field of mechanical technology are found in machines and equipment of various kinds. The mechanical technology is intended for the modernization of the industry. Modernization has been accomplished by introducing power driven fishing crafts or by motorization on the traditional craft. Modernization of the fishing craft in all the states has been undertaken to enable fishermen to go deeper into the sea and fish for a longer time. This has resulted in increased productivity and consequently increased incomes.

The term mechanized craft is used to refer to craft of any size with any range of power. Mechanized craft broadly can be classified into Out Board Motor (OBM) boats and Inboard Engine (IBE) boats. Different types of gears (nets, hooks and lines) are provided and operated from both OBM and IBE boats. The type of gear operated from the boat is the key factor in determining the nature of fishing and the species of fish to be harvested. Trawler, purse-seine, long liner, etc., are the names given to particular types of IBE boats and basically indicate the type of fishing technique and gear operated. OBM boats are called motorized boats irrespective of the type of gear (Srivastav U. K. et. al: 1986). A large number of fishermen have fitted motors of different H.P. to their crafts. Actually, the ocean contains many species of fishes. So far over 25,000 species of marine fishes and shellfishes have been identified. Over 90 per cent of the marine fishery resources are concentrated in about 10 per cent water above the continental shelf area (Ibid).

There is much greater stock of smaller fish in the ocean than larger fish. Small fishes have the tendency to live in water closer to the shore while large fish tend to move away from the shore. The marine fisheries stock grows fast in temperate waters and regions of turbulence (Kurien: 1978). Hence, to tap these vast resources mechanization has been undertaken.

A study by Kurien and Mathew (1982) enumerates three important characteristics in relation to the mechanization of the fishing craft. They are:

1. There are wide regional variations in the species – mix of the marine resource of the country, which lies in the tropical region.

2. The nutritive value of all species of fish is almost the same but the market demand for different species varies and consequently, results in varying prices for different species. For example, oil sardines fetch a very low price (Re.10 per kg.) while prawns, usually called "pink gold", command a very high price (up to Rs. 90 per kg.).

3. The marine fishery resource is a "common property" and is accessible to all. However, the cost of production (catching) varies directly with the distance of the fishing ground from the shore. Therefore, all boat owners, irrespective of the level of technology of their boats, tend to fish as close to the shore as possible (Kurien and Mathew: 1982). As a result of this, conflict between traditional and mechanized fishermen takes place. Traditional fishermen complain of their nets being destroyed by the trawlers.

RESOURCE FACTOR IN TECHNOLOGICAL CHANGE

The impact of technological changes in general and mechanization in particular may not be the same on all fishermen. The impact of technological changes depends on the level of control over resources and markets, the economic and social background of the beneficiaries and the traditional fishermen, and the nature of technological change in terms of labour displacement (Vandana and Bandyopadhyaya: 1982).

Various new types of mechanized crafts have been introduced through Government policy initiatives with a view to harvesting resources, which were so far beyond the reach of traditional craft. The Government never intended to replace the traditional craft. Most mechanized boats instead of going into deep sea have started fishing in shallow waters for various reasons and compete with the traditional craft for resources. This conflict for fishing activities is very much significant in the case of trawlers and purse-seines (Srivastav U. K. et. al: 1986). The evaluation of the impact of mechanization on fisheries is of fundamental importance for resource allocation and employment policy and for the development of fisheries as an organized industry in the Indian economy (Ibid).

The impact of mechanization may be grouped under two headings namely, positive impact and negative impact.

POSITIVE IMPACT OF MECHANIZATION

Positive impact of mechanization is analyzed under the following heads: (1) Impact on production (2) Impact on productivity (3) Impact on employment (4) Impact on earnings (5) Impact on Profitability (6) Impact on Health and Sanitation and (7) Impact on Conservation and Sustainability.

1. Impact on Production

One of the major objectives of technological change in the primary marine fishing industry of Goa was to enhance production by providing the fishermen with better tools and organization. It was also presumed to minimize the fluctuations in the output of the industry. An attempt is made in this section to study the effects of technological changes on the volume and value of output of the mechanized sector of the industry. The data given below reveals the facts about fish production in Goa.

Table 6.1-Important Variety wise Quantity of Marine Fish Catch (in M. Tonns) during 1961 to 2004

S. No.	Items	1961	1971	1981	1991	1995	2000	2004
1	Mackerels	N.A.	35258	4566	4305	42712	16589	6303
2	Sardines	N.A.	2067	11705	28876	1653	19836	34203
3	Catfish	N.A.	83	495	630	1710	676	1043
4	Shark fish	N.A.	117	172	93	25	981	1305
5	Seer fish	N.A.	108	305	984	917	1398	3478
6	Prawns	N.A.	279	3109	3231	5193	2284	5586
7	Pomfrets	N.A.	N.A.	61	242	1105	681	568
8	Others	N.A.	N.A.	N.A.	37262	28541	22118	31908
	Total	-	37912	20413	75623	81856	64563	84394

Source: Directorate of Fisheries, Panaji, Goa

Graph 6.1: Total Quantity of Marine Fish Catch during 1971—2004.

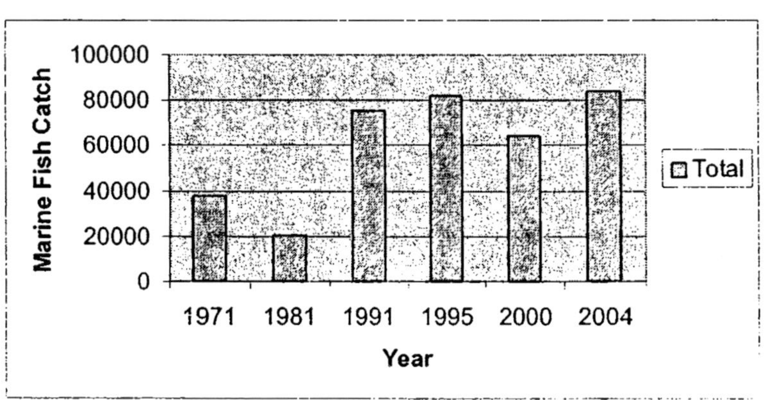

The total quantity of marine fish catch has gradually increased over the period 1971-2004 though in the year 2000 there was a fall in total quantity of fish catch. However, in 2004 again the total quantity of fish increased. The species wise quantity shows a wide variation.

2. Impact on Productivity

A leading objective of technological change in primary marine fishing industry of Goa was to improve the productivity of fishermen and thereby to improve their economic conditions. Neither the directorate of fisheries, nor the fisheries co-operatives have maintained the statistics regarding the fish catch per man-hour after mechanization. Therefore, on the basis of the data collected from the fishermen we can generalize that the catch per man-hour of effort has increased after mechanization. The fishermen revealed that their productivity had increased considerably in the early years of mechanization. However, some trawler owners reported that the productivity is falling because fish stock is depleted in the sea

3. Impact on Employment

An attempt is made here to spell out the effects of technological change on the employment in the primary marine fishing industry of Goa. The technological changes in the primary marine fishing industry of Goa has led to an expansion in the employment opportunities in the activities like boat building, boat-repair, net-making, net-repair, manufacture of wire ropes, winches, floats, sinkers, etc. Trawling, as a new method of fishing, has led to both direct and indirect increases in employment. Directly, it has provided employment to a large number of fishermen in the mechanized boats as labourers, aryamen, khalasis, etc., and indirectly in the storage, processing, transportation and distribution (marketing) of fish.

A large number of migrants are employed as labourers on trawlers and on the motorized crafts. Nearly 90 per cent of the labourers who have migrated from other neighbouring states have got the job opportunities in fishing particularly after mechanization. Even in the leading fish markets of Goa one can find a large number of migrants selling fish. This shows that mechanization has created more employment opportunities.

4. Impact on Earnings of the fishermen

One of the major objectives of technological change and development in the primary marine fishing industry of Goa was to enhance the earnings of the fishermen by providing them with better tools for greater production. Regrettably, the statistical material on this aspect of the change is quite meager to draw any definite conclusion. The discussion with the respondents particularly the motorized and mechanized fishermen revealed that in the initial years of mechanization their earnings were high because the catch was high. In the recent years, though the catch is less, the earnings are still high with the rising prices. It is to be noted that the agents set the price of fish at jetty and they add the value further. So due to lack of proper marketing facilities near jetties the fishermen cannot get the full benefit of rising prices.

A study by Thankappan and Menon (quoted by Korakandy: 1994) observed a phenomenal increase in the earnings of the fishermen in the Indo-Norwegian Project area as a result of mechanization and other improvements. Some trawler owners in Goa expressed satisfaction over their earnings while others said that they are just carrying on the business but there is no much noticeable profit.

Table- 6.2 - Per day gross income of the respondents

Catch (in terms)	Traditional		Motorized		Mechanized		Total	
of Rs.)	No.	%	No.	%	No.	%	No.	%
Less than 2000	61	65.6	29	20.3	-	-	90	22.5
2,000- 4,000	32	34.4	35	24.5	-	-	67	16.8
4,000- 6,000	-	-	32	22.4	28	17	60	15
6,000- 8,000	-	-	28	19.5	36	22	63	15.9
8,000- 10,000	-	-	19	13.3	59	36	78	19.5
More than 10,000	-	-	-	-	41	25	41	10.3
Total	93	100	143	100	164	100	400	100

Graph 6.2: Per Day Gross Income of the Respondents.

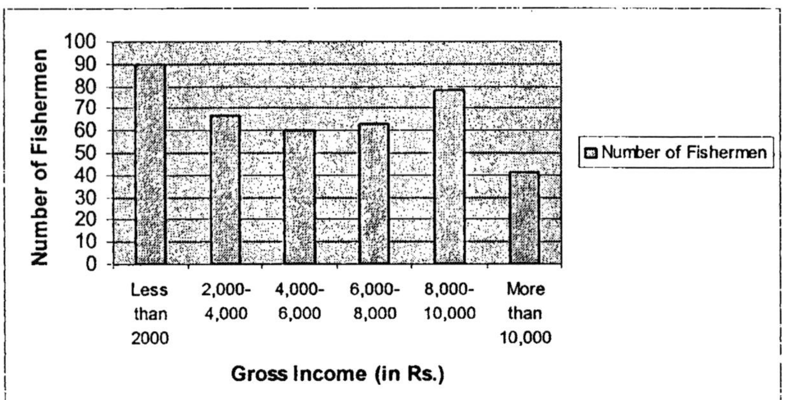

The above data reveals per day gross income of the fishermen. Nearly 23 per cent fishermen earn less than 2,000 rupees a day. Around 17 per cent fishermen earn between Rs. 2,000 to 4,000 rupees. There are very few fishermen who earn more than Rs.10,000. All of them belong to the mechanized section. Per day gross income of the traditional fishermen is not more than Rs. 4,000. This shows that the mechanization has enabled the fishermen to earn more.

5. Impact on Profitability

It has been noted that economic viability is the hallmark for adoption of any new technology. The economic viability of most innovations in fishing techniques in Goa were, however, not rigorously ascertained prior to their introduction as those innovations were carried out simultaneously with research and development efforts by Government support.

During the field survey, we felt that many fishermen were reluctant to give the data about profit earned by them. Most of them were either not sure of the amount of profit made by them or they were scared of income tax being made applicable to them. However, from the type of houses, their assets and vehicles owned by them we can infer that the trawler owners make a considerable amount of profit. The table 6.3 depicts the net profit earned by the fishermen per day.

Table- 6.3-Net profit earned by the respondents per day

Amount of profit	Traditional		Motorized		Mechanized		Total	
(in Rs.)	No.	%	No.	%	No.	%	No.	%
Less than 500	88	94.6	61	43	28	17	106	26.5
501- 1000	05	5.4	47	33	71	43.3	142	35.5
1001-2000	-	-	35	24	38	23.2	97	24.0
2001 and above	-	-	-	-	27	16.5	55	14.0
Total	93	100	143	100	164	100	400	100

Graph 6.3: Net profit Earned by the respondent per Day

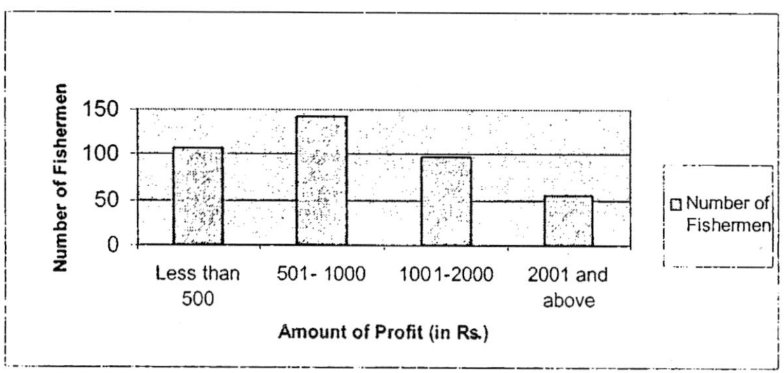

It is evident from the above table that the gross profit earned by the traditional fishermen is less than that of the motorized fishermen and the profit of the trawler owners, i.e., the mechanized fishermen is more than both the traditional and the motorized fishermen.

6. Impact on Housing

A major objective of technological change in the primary marine fishing industry of Goa was to improve the socio-economic conditions of the fishermen who are economically and socially backward. The field survey revealed that the government of Goa has not taken up any measures for housing of the fishermen. The traditional fishermen dwell in poor housing conditions. The motorized fishermen have comparatively well furnished houses. While most of the trawler owners stay in posh bungalows. Most of them have either modified their ancestral houses or have built new concrete houses.

7. Impact on Conservation

The declining catches of the traditional sector has prompted the traditional fishermen to demand conservation of the fishery resources of the state. The fall in the output of the traditional sector was attributed by the traditional fishermen to the introduction of bottom trawling in the state which, they felt, has also resulted in the destruction of juvenile fishes and larvae of many of the species which they were catching. The introduction of purse-seining in the state also had led to a further fall in their catches. Since 1979, the traditional fishermen have been following an agitational path for the conservation of the fishery resources and to safeguard their interests (Korakandy: 1994)

8. Impact of Technology on Fishermen's Income

The effect on the fishermen's income can be gauged from a comparison of returns following the introduction of the fishing technology. Askari and Cummings (quoted in: Harriri Khaled: 1985) used following formula to assess the impact of land reform (i.e., giving land to farmers) on the income of the recipient farmer. In this study we tried to use the same formula to assess the impact of mechanization on the fishermen's income.

$$\Delta\text{"}Y_t = P_t\, Q_{t-}\, It - R_t - Y_{t-i}$$

where $\Delta\text{"}Y_t$ = Change in income due to the introduction of the new technology

P_t = Prices of fish and other marketable species at time *t* after the acquisitions of the fishing technology.

Q_t = Fishery output at time *t*

I_t = Cost of non Labour inputs into fishing

R_t = Annual payments to the creditor and

Yt_{-i} = Income before the acquisition (that is *I* years ago)

The inclusion of R_t presents methodological difficulties. If the transfer of the boat and/or gear to the fisherman was at market value, then the incorporation of R_t is incorrect.

NEGATIVE IMPACT OF MECHANIZATION

1. Resource Depletion

The relatively flat figures for annual catches over two decades raise doubts in the minds of the people about the marine resource depletion.

It is argued with some justification that 'natural factors' play a role in determining marine stocks. For instance, periodic changes in ocean currents and water temperatures, the oxygen content, etc., all result in resource fluctuation. But history testifies that these changes tend to 'even out' over a period of time and the Goan experience over the years under review would seem to reinforce this.

It is also true that marine pollution affects resources adversely. The release of industrial effluents and wastes, mining rejects, unprocessed sewage and toxic materials cause substantial damage to fishery resource. And scientific studies carried out in Goa's waters provide evidence that the estuarine spawning grounds are under cognizable threat. Yet, at this point of time, the overall effect of pollution on fish catches (except clams) may be considered marginal.

All things taken into account, it would appear that the ill-advised adoption of supposedly 'modern technology' is the major cause of problems plaguing Goa's fishing industry today. More precisely, the deployment of this technology to exploit marine resources beyond sustainable limits has resulted in their steady depletion, almost to a point from which a healthy recovery could only be miraculous. Scientific opinion buttresses the fear.

Parulekar and Dalal (1979) estimated that the maximum sustainable yield (MSY) of the marine fisheries off the coast of Goa is 70,295 tonnes per annum – 41,760 tonnes 'pelagic' and 28,535 tonnes 'demersal' (the estimate is subject to natural variations). On the basis of this study, 60,000 tonnes per annum would be the maximum harvestable yield. The annual catch has been much more than 60,000 for the years 1991-2004. (Table 2.2)

Actually, some people say that the marine resources of Goa are underexploited and the policy of increasing mechanization is sound and justified. They also feel that the demand of the traditional fishermen for conserving fish resources is premature and irrelevant. But the reality is somewhat different.

The MSY, as calculated by Parulekar and Dalal, refers to all fishery resources off the coast of Goa, including the deep sea. And Goa's catches are all fairly close to shore.

One can divide the entire fishing area into three sectors:

1. The area between 0-10m depth, reserved for the traditional fisherman;
2. The 10-15m, depth zone which can be exploited by the existing trawlers and purse-seines; and,
3. The area marked for deep sea fishing.

The major problem in Goa is that it is essentially sector one which is being viciously overexploited, to the detriment of the local ecology. The traditional fishermen in their mechanized (Out Board Motor) and non-mechanized country craft (about 3000 in number) cannot venture very far out to sea; so in their case, fishing close to shore is perfectly justified and legal. But the trawlers fishing closer to the shore cannot be justified. In the year 2005, there is another technology started by the trawler owners which is called as Bull-pulling trawlers. Two trawlers are joined together and a big net is spread in between. When these trawlers go sailing, all the fish in between the two trawlers is trapped in the net.

The main argument against increasing the number of trawlers and purse-seiners is that they are not the appropriate technology form for optimizing fish yield. There are now about 1134 trawlers, and according to Alvares (2002) there is no conceivable way in which such a large number of vessels can be supported by a coast 105 km in length.

Rapid adoption of new production technologies in a context of resource scarcity has become a topic of concern for the people involved in fisheries.

2. Concentration of Production Capacity in few hands

The capital-intensive nature of most new fishing technologies has led to increasing concentration of production capacity in the hands of relatively few individuals. As a result there has been (1) a de facto reallocation of scarce resources favouring those who control the most powerful technologies and (2) the creation of a dualistic industry structure characterized by important difference in technology capitalization and basic orientation of the fishing enterprise itself.

3. Excess Fishing Effort

Given the state of our knowledge regarding population dynamics of tropical multi-species fisheries, and the common absence of data on existing levels of fishing effort, we cannot with precision establish at what point levels of fishing effort become excessive. As fishing effort increases, there may be differential impacts on the resource, with the result that certain species may decline in abundance while others may increase (Pauly 1979). Under these circumstances, increasing levels of fishing effort beyond some mythical point identified as maximum sustainable yield (MSY) does not automatically result in declining catch but may instead result in a changing species composition which may (or may not) have a higher economic value.

Recognizing these limits to our knowledge, it is generally accepted that levels of fishing effort can increase to the point where most or all commercially valuable species suffer population declines. This resource depletion may be defined as the point where levels of fishing efforts have become excessive. Levels of fishing effort which exceed some estimate of MSY are not by definition "excessive' unless they result in resource depletion. The general feeling of the fishermen is that the fish resource is depleted and the high value fish is becoming extinct.

4. Competition between small-scale and large scale fishermen

Competition between small and large-scale fishermen is of two-kinds: competition over finite marine resources and competition over scarce development resources.

5. Competition over marine resources

Competition between small scale and large scale fishermen occurs when both types of fishermen exploit the same fishing grounds and/ or the same species of fish.

The small-scale fishermen are restricted to coastal waters by their simple fishing technologies. Trawlers in particular are attracted into coastal waters by the relative abundance of high-valued shrimp. In the process of trawling for shrimp, large quantities of other organisms also are captured, including a high proportion of sexually immature but commercially valuable fish and shrimp (Azhar: 1980). Despite eroding catch-per-unit effort ratios, excessive levels of fishing effort goes on due to the high prices paid for shrimp in the world market. This, in turn, has led to serious declines in catches and incomes among large numbers of small-scale fishermen (Bailey: 1982, 1985, 1987; Panayotou: 1980; Smith: 1979).

6. Resource depletion and threats to sustainable harvests

Most important fisheries are being exploited at or near the level where there exists a clear threat of resource depletion. For societies as heavily dependent on fisheries resources for nutritional and economic well-being, this doubtlessly is the most serious negative social consequence of excessive fishing effort. In broad terms, excessive fishing effort represents a misallocation of human and capital resources resulting in (1) a diminution of economic returns to these two factors of production, (2) declining harvests because of resource depletion, and (3) consequent losses in income, employment opportunities, and supply of fisheries products.

7. Gear conflict

Competition for a dwindling resource base has led to serious gear conflict, especially between small-scale fishermen and trawlers. Trawlers not only compete effectively against small-scale fishermen, but because of their active mode of operation, they frequently damage or destroy more passive small-scale gear. This is a particular problem at night, a time favoured by trawlers operators as it is then that shrimp are most active and easily caught. As an added incentive, trawlers operating illegally in coastal waters are less likely to be apprehended at night (Panayotu

1980:44). Majority of the traditional fishermen we interviewed complained about this problem.

Destruction of small-scale gear by commercial trawlers has also been reported in other countries like in Philippines (Bailey 1982; Smith 1979), Malaysia (Bailey 1983; Gibbons 1976; Smith 1979), and Indonesia (Sardjono). Damage or destruction of small-scale fishing gear caused by trawlers results in serious economic losses and is a continuous threat to the life and livelihood of many small-scale fishermen.

8. Social Conflict

Conflict between competing users is an inevitable result under conditions of excessive fishing effort. Technological advantages are not only a matter of economic class but are also frequently related to ethnic divisions within the region.

9. Increased use of destructive fishing techniques

As excessive levels of fishing effort lead to resource depletion, fishermen are forced to adopt increasingly fine meshed nets, bull-pulling trawlers or other destructive fishing methods, which meet their short term needs at the expense of long-term interests in resource sustainability. This response to resource depletion is most likely to harm the small-scale fishermen who have no alternative employment opportunities. Under these conditions, which are common throughout the state, fishermen have no economic alternative to exploiting undersized or juvenile recruits before they have reached optimum economic value.

10. Impacts on food supply and distribution channels

Fish provides the single most important source of high quality animal protein, and the only affordable source for the poor. Demand for fisheries products within the state is increasing due both to natural population growth rates and to increasing perfect competition per capita demand by urban and middle-class consumers, who have benefited most from national economic development.

As levels of fishing efforts increase to the point where resources begin to be depleted, fish becomes an increasingly valuable commodity in domestic markets. As prices rise, the ability of the poorest strata within society to purchase fish is diminished, with the predictable result being increased protein malnutrition. This is the general trend found in the state. Poor people cannot afford to consume fish except in case of peak season, i.e., from September to November. Poor people can buy, at the most, low value fish like Sardines, ribbonfish, etc. They cannot buy fish during the lean season i.e. from December to May as fish is expensive during that period.

The growth of large-scale fisheries has led to a major shift in the centre of fishing activities from small coastal communities to urban fishing ports equipped to handle larger boats. The relative affluence of urban compared to rural populations has increased the importance of urban markets as centres of demand for fisheries products. The expansion of large-scale fisheries not only has contributed significantly to excessive levels of fishing efforts, but also increased the importance of urban fishing ports as centres of supply for fisheries products. Urban consumers, in particular, have benefited from this development. Rural consumers, however, have been adversely affected wherever an increasing proportion of the catch is landed at urban ports, because from there distribution patterns tend towards other urban markets rather than to rural areas.

11. Changes in the social relations of production

Large-scale fishermen enjoy a number of important advantages which contribute to their increasing importance within the fisheries sector. The greater productivity of large-scale compared to small-scale fishing units provides an important advantage in competing for a dwindling resource base. Direct and indirect subsidies give this technical edge an economic dimension of apparent efficiency, allowing commercial fishermen to compete for a dwindling resource on advantageous terms with small-scale fishermen (Smith and Mines 1982).

This combination of technological and economic advantage has resulted in the concentration of fishing power in the hands of relatively few individuals. With the concentration of fishing power, concentration of economic power has emerged. Labour is in relatively abundant supply compared to capital, and owners of capital-intensive fishing units are able to dictate the terms of employment (Villafuerte and Bailey 1982).

The examination of sharing systems provides valuable insights into social and economic relationships within the fisheries sector. The manner in which the proceeds from the sale of the catch are distributed among owners and non-owning crewmen reflects the respective value place on capital and labour as factors of production. In small-scale fisheries, it is common to find owners taking an active role in fishing. Analysis of sharing systems reveals that clear distinctions are drawn between capital, labor and management, but as factors of production among small-scale fishermen, these frequently are combined in the role of owner-operator. Within the small-scale sub sector, sharing systems frequently are based on more than economic calculation (Aminah and Widjayanti 1980; Bailey in press; Villafuuerte and Bailey 1982).

Large-scale fisheries enterprises are operated quite differently. Owners provide capital and on-shore management, paying particular attention to marketing, but leave management of actual fishing operations to a hired captain (Bailey 1983; Villafuerte and Bailey 1982). This captain is responsible for hiring and firing of the crew, who have little contact with the owner. Thus, the roles of investor, manager and worker are clearly differentiated. The result is a separation between owners and crewmen along lines of economic class interests. The separation of ownership, management and labour serves to lessen owners' social obligations, a buffer which owners are careful to maintain (Villafuuerte and Bailey 1982). This study also confirms similar complaints from the labourers.

12. Resource allocation and distributive justice

Fisheries development efforts which focus exclusively on production-oriented technologies raise serious ethical problems associated with distributive justice (Bailey et al. 1986). The capital-intensive nature of purse seiners, trawlers, and other highly effective modern fishing technologies precludes all but the wealthy few from benefiting from this form of development. By promoting the use of highly productive technologies without simultaneously strengthening institutional capacities to manage and allocate finite resources among competing users, national and international development agencies are contributing to structural problems and policy distortions which result in excessive levels of fishing effort and negatively affect the catches, incomes and standards of living of the majority of those employed in the fisheries sector.

SOME NEGATIVE FACTS ABOUT GOA'S FISHERY SECTOR

The scenario in the state of Goa is that as in other parts of the country and the world, the Government encouraged the traditional fishermen to adopt new technology. Introduction of technologically advanced vessels like trawlers and purse seines, for commercial purposes in the last three decades has drastically affected the coastal traditional fishermen's activity in Goa. The mechanized vessels are engaged in trawling which is ecologically hazardous, as they fish in coastal waters and have destroyed the seeding grounds of shrimps, mackerels, and other small crustacean families that cliché closely to coastal waters for food, breeding and living in harmony. They have exploited shallow water fishing grounds where the traditional Ramponkar has been fishing for generations. Paradoxically, these same trawlers have been the prime cause of marine pollution through oil spills and wastes. Invasion of these vehicles has led to higher ratios of traditional Ramponkar community being thrown out of their traditional livelihood.

Today, the traditional Ramponkar communities of Goa's three talukas live on the threshold of being economically strangulated. Fishermen from Canacona, Salcete and Bardex talukas have already become victims of rampant "Trawling". Technologically advanced "Trawling" has not only eroded Goa's estuaries and fish nurseries, but this encroachment has affected fish yield each year for over a decade. This has now led to Goa being declared a fish famine zone by the Central Fisheries Board. There has been a drastic change in the composition of fishing crafts in Goa. The number of mechanized boats increased over the years.

Table 6.4 Number of Fishing Crafts in Goa and their increase during 1960-61 to 2003-04

Year	Mechanized Boats	Net Increase in Mechanized Crafts	Country Crafts	Net Increase in country.	Total fleet
1960-61	4 (0.10)	-	4125 (99.90)	-	4129
1970-71	110 (2.96)	106	3600 (97.04)	- 525	3710
1980-81	189 (6.14)	86	2887 (93.86)	- 713	3085
1985-86	600 (19.67)	411	2450 (74.36)	- 437	3050
1990-91	707 (25.64)	107	2050 (74.36)	- 400	2757
1995-96	880 (30.77)	173	1980 (69.23)	- 70	2860
1997-98	1056 (36.46)	176	1840 (63.54)	- 140	2896
1998-99	1092 (36.42)	36	1890 (63.38)	50	2982
1999-00	1092 (33.23)	-	2194 (66.77)	304	3286
2000-01	1128 (36.49)	42	1963 (63.51)	- 231	3091
2001-02	1128 (36.49)	-	1963 (63.51)	-	3091
2002-03	1134 (36.6)	6	1963 (63.4)	-	3097
2003-04	1134 (36.6)	-	1963 (63.4)	-	3097

Source: Directorate of Fisheries, Government of Goa

Note: Figures in parenthesis indicate percentages of the total fleet.

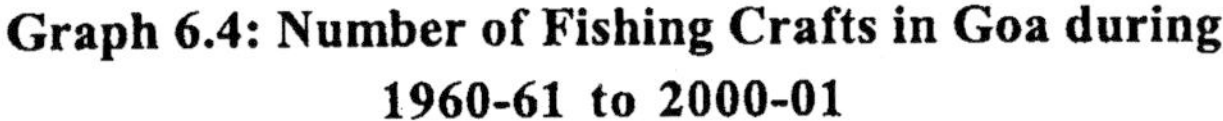

Graph 6.4: Number of Fishing Crafts in Goa during 1960-61 to 2000-01

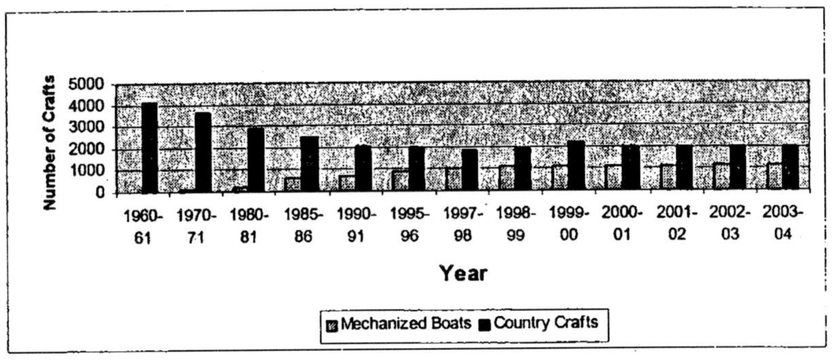

Data regarding the number of fishing crafts in Goa over the years from 1960-61 to 2003-04 shows that in 1960-61 the number of mechanized boats were less than 1 per cent while country crafts accounted for more than 99 per cent of the total fleet.

The figures after 1960-61 show that the percentage of mechanized boats has been steadily increasing while the percentage of country crafts has been decreasing. In contrast to the 1960-61 figures, the 2000-01 figures show that the mechanized boats contribute 36.49 per cent while the country crafts account for 63.51 per cent of the total fleet.

Another revealing fact is that from 1960-61 to 2003-04 the number of total fleet has decreased from 4129 in 1960-61 to 3097 in 2000-01. This fall is due to the fall in the country crafts. This reveals that after mechanization the number of traditional country crafts has declined because they do not find fishing a remunerative business, some of them have switched on to mechanization and some are working as labourers on the mechanized boats. Some others buy fish from the wholesalers and sell in the retail market.

The discussion with the fishermen reveals the fact that as the mechanization was rapid and the numbers of boats were more than required for a total length of 105 kms, the fish catch is not increasing. It has remained more or less stable or has increased only marginally.

The State of Goa has 1134 registered trawlers and this is much above the saturation point. The Food and Agriculture Organization

(FAO) of the United Nations has recommended 30 trawlers per 10 kms of coastline (Barbosa: 2002). Given that Goa has 105 kms of coastline the number of trawlers should have been around 315. But instead there are 1134 trawlers.

The number of trawlers increased with the new schemes of the Government of Goa. It also developed fishing jetties at various points in the State. Without paying any attention to the existing fishing facts and experiences of other countries, the Government took steps to promote mechanized fishing. Interestingly, in 2001 the Government realized the need to stop the registration of the new trawlers. In reality, the State of Goa has more than 4 times trawlers as compared to the recommended formula of the FAO. Besides, the Government has provided fish finders to the trawler owners. Many fishermen feel that the fish finders will pose a threat to sustainable development of fisheries.

An important finding of this study is that though the number of trawlers has increased, the fish catch is not increasing. From 17000 tonns in 1961-62, it reached an all time high in 1993 when the catch reached its peak at 97014 million tonns after that it never touched that level. Instead it showed a decline for some time and then a gradual but marginal increase.

Table 6.5 Total Fish Catch in Goa from 1961 to 2002

Year	Production in Tonns	Net Increase in Production
1	2	3
1961-62	17,000	–
1970-71	36,616	19616
1980-81	25,715	-1090
1985-86	27,711	1996
1990-91	56,225	28514
1992-93	97,014	40789
1995-96	85,418	-11596

1	2	3
1997-98	94,547	9129
1998-99	70,710	-23837
1999-00	63,440	-7270
2000-01	69386	4693
2001-02	73,135	-3749
2002-03	83756	10621
2003-04	84394	638

Sources:

1. D'souza: 2002: Goa Today November 2002. pp 10-14.

2. Directorate of Fisheries.

For example, since 1992-93 the quantity of fish catch declined continuously from 97,014 million tonns to 63,440 million tonns in 1999-2000. Since 2000-2001, it gradually started rising and the total fish catch in Goa was 84,394 million tonns.

As the demand remains more or less the same and hence, the prices tend to rise with the falling supply of fish, the fishermen make up for the costs they incur.

The traditional fishermen complained that the noise of the mechanized crafts frightens off the fish and correspondingly reduces the catch. The bottom part of trawlers cause damage to fish eggs and larvae and disturb nursery grounds of fish.

A major problem resulting out of mechanization is increasing inequalities in inter-personal income level that is between boat owners and traditional fishermen and between workers in the mechanized and non-mechanized sectors. The growing income difference has widened the gap between the rich and poor fishermen, creating the two classes namely; mechanized and non-mechanized fishermen.

Mechanization has also disturbed the fish ecology. The trawler owners throw the low value catch back into the sea. The field study revealed that nearly 50 per cent of the trawler owners in our study throw the less value fish back into the sea. In the process of catching and throwing some fish die. The dead fish if thrown back in the sea disturbs the fish ecology.

Table 6.6– Respondents' opinion on availability of fish resource in sea

Availability of Fish	Number	Per cent
Increasing	3	0.7
Stable	8	2
Depleting	389	97.3
Total	400	100

The respondents were asked to mention the names of fishes they feel are depleting in the sea. Their responses are presented in Table 6.7. According to them the high value fish namely pomfrets, kingfish and prawns are depleting.

Table 6.7 Names of depleting fishes

Sr. No.	Names of Depleting fishes	Number	Per cent
1	Mackerel	52	13
2	Shark	22	5.5
3	Prawns	53	13.3
4	Pomfrets	84	21
5	King fish	98	24.5
6	Ribbonfish	26	6.5
7	Squids	13	3.3
8	All fishes	29	7.3
9	Catfish	12	3
10	Velim	11	2.6
	Total	400	100

Another impact of mechanization is that the fishermen have started reducing the mesh size which catches even the small fish. This will further lead to depletion of fish resource. It is fact that the mesh size of the nets has reduced over the years because there are no mature shoals of fish. With an intention to fill their nets the fishermen go for the smaller fish, or rather baby fish. The fishermen use a mesh size of even 16 mm and 18 mm, which is not allowed otherwise.

On the basis of the above analysis we can conclude here that the negative impact of mechanization is more on the fishery sector compared to the positive impact. Therefore, the Government should not allow the boat owners to build new boats in place of the old boats. Though, the Government has stopped giving licenses for new boats, it has done nothing to stop the worn out boats.

The study revealed that some fishermen buy the registration numbers of the old boats from the trawler owners of old boats and in place of the same, they build the new boats. This should be stopped for achieving sustainable development of the fishery sector.

CHAPTER VII

CONFLICT: THE POLITICS OF ENVIRONMENT AND LIVELIHOOD

Sustainable development is at the core of all fishery management strategies. Development should go in pace with the availability of resources as to avoid the consequences of haphazard growth. The bounty of nature is unfathomable, but a century of wanton and reckless exploitation and harvesting for man's greed and lust has dwindled the otherwise inexhaustible resources. Man has taken nature for granted little realizing that over-fishing will always lead to extinction of some species of fish and ultimately to fish famines.

Before liberation fishing in Goa was at subsistence level. Fishing was done in the traditional way. It was only for consumption. But in the post liberation period the benefits of mechanization were extended to Goa. The Government of Goa in its ambitious programme of mechanizing the fishing sector offered subsidies to the fishermen to go for the new technology. But the artisan fishermen because of his ignorance of new technology and lack of finance did not take the offer made to him. This gave opportunity to non-fishing communities to enter into the fishing arena. The lucrative fishing industry also attracted politicians. They became trawler owners directly or in benami names (GRE Souvenir 1975-1995). It was then that the artisanal fishermen realized that the new competitors were at more advantaged position because of their advanced technology. The mechanized trawlers and purse seines instead of going in the wider sea begin to stray into areas which had been traditionally that of the artisan fishermen, as special species of prawns were available there. The artisan fishermen felt threatened and displaced and looked to the government for support. They criticized what they called the "thoughtless mechanization" that spurred growth in the industry but at the cost of long-term sustainability of the industry it self (Ibid). As no democratic method would work in 1975, the artisan fishermen like Pagelkar, Ramponkars organized themselves under the name of Goencho Ramponkarncho Ekvott (GRE).

In 1976, the GRE resorted to chain hunger strike of 380 days and submitted a memorandum to the then Chief Minister Mrs. Shashikala Kakodkar to protect its interests. The government gave a Nelson eye, accusing the artisan fishermen of sabotaging progress and development. The artisan fishermen accused the mechanized boats of fishing very close to the shore depriving them of their livelihood. The GRE claimed that the vested interest in the elected legislature instead of protecting the interests of the people was going against it.

The GRE held that there was no law to protect the artisan fishermen. It claimed that the 1897 Act which was designed to ban the use of dynamite provided a ban on trawling and pursening within the depth of 5 fathoms, which was two kilometres from the shore, an area insufficient for the traditional communities (Ibid).

The struggle of artisan fishermen continued even during the emergency of 1975. In 1978 the Government's 5 fathom rule was challenged by the trawler owners in the Judicial Commissioner's Court, presided by late Justice Tito Menezes. Justice Menezes gave the ruling that rules were bad in law and that Government could not prevent one section of the fishermen from fishing and allow others to fish in the given area (Ibid).

The artisan fishermen intensified their pressure on government. They organized morchas, agitations and courted arrests in support of their demand that government should make a law to ban trawling or mechanized fishing within 5 kms from the shore. They demanded a ban on night trawling and pursening and fishing during monsoon season. But the Government which was then headed by Mrs. Shashikala Kakodkar reacted by branding the fishermen's movement as anti-national and its leaders as trouble makers (Ibid).

It was then that 80,000 fishermen and their leaders decided to organize the fishing community at all India level as the, adverse consequences of mechanized fishing were felt all over India. In 1978, Mr. Xavier Pinto contacted fishermen leaders in Karnataka, Maharashtra, Kerala, Tamil Nadu, Andhra Pradesh and Orissa. The result was the formation of the National Forum for Catamaran and Country Boat Fishermen's Rights and Marine Wealth [the present National Fisher Workers Forum (NFF) (Ibid.) Shri Mathany Saldanha was elected as the chairman of the organization. The NFF demanded a national law to protect ecology and to ban mechanized fishing within 5 kms. It approached Morarji Desai's Government and the members of Parliament to pass such legislation. It was in 1980 that the newly elected Government of Mrs. Indira Gandhi circulated to all States a draft of the Marine Regulation Bill, to be enacted in their respective Legislative Assemblies as fisheries was a State subject (Ibid).

In 1980, the first Congress Government was formed in Goa with Dr. Wilfred D'souza as the Fishery Minister. Under the pressure from the traditional fishermen the Government passed the Goa, Daman and Diu Marine Fishing Regulation Act, 1980. Mr. Mathany Saldanha claims that the Goa Government modified the Central Regulation Bill to satisfy the vested interest of the elected legislators (Mathany Saldanha Interview: 2000).

The Goa, Daman and Diu Marine Fishing Regulation Act, 1980 (Act 3 of 1981), though not fully welcomed by the GRE, it accepted their long demand of regulating fishing by fishing vessels in the sea along the coastline of the Union Territory of Goa, Daman and Diu. Clause 4 in the Chapter II gave power to regulate, restrict or prohibit certain matters within the specified area. Under Chapter 1, 2 (h) "Specified area" means such area in the sea along the entire coastline of the Union Territory or such portion of it but not beyond the territorial waters as may be specified by the Government by notification in the Official Gazette (see The Goa Daman and Diu Marine Fishing Regulation Act, 1980).

In keeping with the Act, the Government issued a Notification 2-1-81-FSH (III). In exercise of the power conferred by clause (h) of section 2 of the Marine Fishing Regulation Act 1980 (3 of 1981), the Government of Goa, Daman and Diu specifies an area of 5 kms from the coastline in the sea along the entire coastline of the Union Territory as a specified area for the purpose of the said Act (Government Gazette Series /No. 15,9-7-1981). In another Order 2-1-81-FSH (IV), in exercise of the powers conferred in sub-section (1) of section 4 of the Marine Fishing Regulation Act of 1980, the Government of Goa Daman and Diu having regard to conserve fish and to regulate fishing on a scientific basis, prohibits fishing with mechanized vessels in a specified area declared as such under Government Notification No. 2-1-81-FSH (III) dated 7-7-1981.

In 1981, S. L. P was filed before the Supreme Court by All Goa Boat Owners Association challenging the constitutionality of the Goa Daman and Diu Marine Fishing Regulation Act, 1980. The apex Court passed an Interim Order dated 9-10-1981 by which it reduced the fishing area from 5 km to 2 ½ kms and put a total ban on night fishing (Fishing File, Goa Foundation). In 1993, the Supreme Court passed an Order directing State Government to make a thorough study of the case within one year with the help of the scientists and fishermen and to inform the Court of its findings. In 1994, after hearing from the State Government, the Supreme Court passed an order banning mechanized fishing within 5 kms from the shore and empowered the State to make rules to conserve marine ecology by banning mechanized fishing during monsoons (Fish File, Goa Foundation).

The conflict between the traditional fishermen and the trawler owners continued. The GRE accused the boat owners of violating the 5 km ban. GRE warned the Government that the depleting fish resources would cause a fish famine like the one in 1990.

The Goa Government assured that it would act positively. In a Notification 15th February, 1995 the Government stated "In exercise of the powers conferred by sub-section (1) of section 4 of the Goa, Daman and Diu Marine Fishing Regulation Act, 1980 (Act 3 of 1981) the Government of Goa, having regard to the need to conserve fish and to regulate fishing on a scientific basis, hereby prohibits mechanized fishing by means of trawlnets and purse-seine nets along the coast of Goa for the period from 1st June to 31st August (2-1-85-FSH 15 Feb 1995). This Notification was superseded by another Notification on 17th July 1995 which stated: In exercise of the powers conferred by sub- section (1) section 4 of the Goa

Daman and Diu Marine Fishing Regulation Act, 1980 (Act 3 of 1981), the Government of Goa, having regard to the need to conserve fish and to regulate fishing on a scientific basis, hereby prohibits mechanized fishing by means of trawl nets and purse-seine nets along the coast of Goa for the period from 1st June to 24th July every year (Official Gazette Series No. 15 dated 18-7-1995).

The Fisheries Department claimed that the reduction of the period from 90 to 54 days was done on a representation received from the mechanized fishing vessels operators who found the 90 day period too harsh as it deprived them of their earnings as during this period the trawlers were idle and even then during this time they had to pay the wages of the crew engaged on the vessels. The Fisheries Department further maintained that this period was reduced after a consensus was reached at the joint meeting of the traditional fishermen and the mechanized fishing vessels operators on 23rd June 1995 in the T. B. Cunha Hall, Panjim. On 8th April 1999, Government of Goa issued yet another Notification:- In exercise of the powers conferred by sub-section (1) of section 4 of the Goa, Daman and Diu Marine Fishing Regulation Act, 1980 (Act 3 of 1981), the Government of Goa, having regard to the need to conserve fish and to regulate fishing on a scientific basis hereby prohibits fishing by means of trawl, purse-seine, gill nets, etc., with mechanized fishing vessels, and motorized fishing craft, in the specified area along and on the sea coast of the State of Goa declared as such vide Government Notification No. 2-1-81 FSH (III) dated: 7-7-1981 throughout the year (Official Gazette Series 1 No. 5 29 April, 1999).

The Government of Goa hereby further prohibits mechanized fishing also beyond the specified area declared under Government notification No. 2-1-81 FSH (III) dated: 7-7-1981, and up to the territorial waters limits along the sea coast of the State of Goa for the period from 1st June to 24th July every year and also prohibits catching of juveniles of fishes such as Mackerels, Sardines, etc, during the period of monsoon as well as in the fair season, with a view to conserve fish and regulate fishing on a scientific basis. The period of monsoon meant the period commencing on 1st June and ending on 24th July and fair season meant the period commencing on 25th July and ending on 31st May of the succeeding year (Ibid).

Directorate of Fisheries claims that this Notification introduced additional safeguards to maintain marine ecology by prohibiting mechanized fishing beyond specified area and extending the ban up to the territorial limits which is 22 kms along the sea coast of the State of Goa for the period from 1st June to 24th July every year. Moreover, it had banned the killing of juveniles of fishes. The Fisheries Department further claimed that the policy decision of a ban of 54 days was taken after the matter was closely considered, taking relevant aspects including the availability of fish and local fishing operations.

Too many Notifications and inconsistency shown by the Government as regard the ban period during the breeding season of the fish created an uneasiness and grave concern among the general public. Critics claimed that it only exposed the sincerity and wishes of the Government in conserving the fish resources. It also confirmed people's suspicion that the Government was dancing to the tune of the rich trawler owners. As July 2000 was approaching there was the public apprehension that the Government might go for a further reduction of the ban period. On 5-7-2000 GRE warned the Fishery Minister and the Goa Government against a further reduction of the ban period, on the contrary it suggested the extension of the ban up to 31st August (Fishing File, Goa Institute of Research, Panjim).

Public anxiety found its way in the corridors of the Bombay High Court, Panjim Bench. On 5 July 2000 in a PIL writ petition No 212 of 2000 Shri Marcos Santan Otilio Pinto De Santan of Divar raised certain issues before the Bombay High Court. The petitioner claimed that the months of June to August are the breeding season and the fishing activity is banned . But despite the ban fishing is being carried out and to add insult to the injury not with standing full knowledge of the negative effects on the ecological balance, Government had reduced the period of ban on fishing from 90 days to 54 days. The petitioner claimed that this was done without proper application of mind and without any scientific data. He warned that a strong lobby was parading in the corridors of the Secretariat working overtime on the modalities for a further reduction of the already reduced period (Goa Foundation, Fishing File). The petitioner claimed that action of the Government not only created "an imbalance in marine ecology" but it has consequential effects by pollution, caused only to satisfy the greed of these poachers with total disregard to law, ethics and morality (Ibid).

The petitioner further held that Government was duty bound to "protect ecology and also the marine life but unfortunately Government and the concerned Department were held ransom to the lobby which included certain elected representatives who had a vested interest for their own selfish motives with least concern for the preservation of the fundamental aspect of nature". The petitioner sort Court intervention to "stop the destruction and the rape of ecology which in the long run would affect an already fragile ecology and distressing economy." The petitioner pleaded Court intervention to "salvage the situation" and to give necessary direction for the rectification of these illegal activities and to ensure control and protection which would benefit not only State at large but also generations to come (Fishing File, Goa Foundation).

Taking the cognizance of the letter on 6th July 2000, the High Court Judges Justice Rebello and Justice V.C. Daga maintained that petition raised question of fundamental right to life of all citizens of the State who through centuries have been consuming fish which is the cheap source of protein and also the right of traditional fishing community. They called on the Director of NIO to file affidavit based on the material available as to the breeding season of fish of the Goa coast and the estuaries areas and period for which ban should be imposed to protect the breeding season (Ibid). The Court asked the Director of Fisheries to file an affidavit providing such information as (a) what was the ban period imposed by the coastal States of Kerala, Karnataka, Gujarat and Maharashtra during the breeding season (b) What measures can be taken to prevent mechanized fishing boats operating in the rivers or in the sea during this period (Ibid). The Court asked Chief Secretary of the State of Goa to file a Notification which had altered the ban to 24th July (Ibid July 2001). Advocate Norma Alvares was appointed as Amicus Curiae to assist the Court in the matter.

The Director of Fisheries in an affidavit of 18 th July 2000 stated that notification of 8th April 1999 imposed additional safeguards which prohibited the catching of the juvenile fish. It argued that those who use trawler nets or carry fishing by gill nets for catching prawns during late July and early August would suffer great disadvantage and financial loss if the ban is extended throughout the monsoon period. He also stated that the monsoon ban from 1st June 31st August was brought down to 1st June 24th July taking into account all relevant aspects and interests and in consultation with the traditional fishermen and the mechanized boat owners. It further stated that the ban imposed in various coastal States of Kerala, Karnataka, Gujarat and Maharashtra.

Kerala 45 days 15-6- to 29-7

Karnataka 92 days 1-6- to 31-8

Maharashtra 62 – 67 days 10-6 to

Narili Puronima or 15 August which ever was earlier

Goa 54 days 1-6 to 24-7

Gujarat had no act (Fishing File, Goa Institute of Research, Panjim).

In a Supplementary affidavit filed on 20th July 2000, the Director of Fisheries argued that the breeding period of Indian Mackerels and Sardines was from May to August and that ban on the species has been imposed for fishing by mechanized trawlers during the entire year up to 5 kms. And during monsoon for a period of 54 days and the ban on fishing of juvenile fishes was during the whole year, as such the peak of the breeding season was well covered (Fishing File, Goa Foundation). His final submission was that this 54 days ban period was adopted after taking into consideration all aspects and it was from 17th July 1995 and continued till date as modified vide notification dated 8-4-99. This was in force nearly for five years (Ibid).

In his affidavit the NIO Director acting through senior scientists Dr. Ansari, Dr. X. N. Verlekar and Shri. R.A. Sreepada, stated that sustainable fishery depends upon successful spawning. It has been documented that fish catch has shown marked improvement after the introduction of the trawl ban. The affidavit further claimed that the available data suggest that the breeding/ spawning behaviour of some species coincide with the south west monsoon. It concluded that based on available published information along the west coast of India a ban on fishing season from 1st June to 31st August could offer a degree of protection to a few fish, like Dhoma fish, croaker, Indian mackerel and Indian oil sardine. It admitted that other species like pomfrets, white prawn, sea crab and cat fish would go completely unprotected (Ibid).

Taking cognizance of the recommendations of NIO the Panjim Bench of the Bombay High Court consisting of Justice R.K. Batta and V.C. Daga on 20th July 2000 gave an interim order. The learned judges agreed that affidavits filed by the Fisheries Department showed that decision to reduce the period of ban from 90 days to 54 days was taken on 17th July, 1995. Although they had asked State of Goa to place before the Court the material, scientific data which was taken into consideration for taking a policy decision regarding reduction of ban from 90 days to 54 days no such data was made available. The judges concluded that it appeared that the policy decision was taken mainly on account of the representations made by the mechanical fishing vessels who pleaded that the ban acted harshly on them. The judges maintained that any policy decision should take into account the breeding season of the fish. In the absence of data relating to breeding season the Court ordered the stay of the notification/ order 2-1-81 – FSH/6 dated 8th April, 1999 and the notification dated 17th July 1995 published in Official Gazette, Government of Goa extraordinary Series I N. 15 dated 18th July, 1995, till 15th August, 2000. The Court concluded "in other words, there shall be complete ban on fishing with reference to the said notifications till 15th August. The Court stated that Government could place before it the marine scientific data, which it had taken into consideration while taking the policy decision of fixing the ban period from June 1st to 24th July. The matter was fixed for consideration for 18 August 2000 (Ibid).

The Interim Order raised a hue and cry in the fishing circles. A delegation of mechanized boat owners led by MLA Jose Philip and MLA Churchill Alemao both trawler owners, met the then Chief Minister Sardinha and expressed their difficulties. The apprehension of the boat owners was understandable; they would have to wait for their 'catch' which would mean losses of income. But what was amazing was the dust and heat raised during the zero hour in the Goa Legislative Assembly. Members cutting across the party lines created a racket over the extension of the ban. Shri. Churchill Alemao was all fury and fire. He accused the Government of playing the mute spectator (Herald July 2000). He argued that the ban of 76 days was detrimental to the interests of the trawler owners who were running huge losses. They would be doomed as the solar shrimp which enabled the trawler owners to survive the remaining part of the year would be lost as it was available between June and mid July. He demanded to know as why the Government had failed to file a caveat in connection with the PIL. He called on the Government to issue a notification to come to the rescue of mechanized vessels as the trawlers from neighbouring State were fishing in Goa's waters and had gone away with huge quantities of fish (Navhind Times July 2000). Shri. Sanjay Bandekar, a MLA from Canacona, himself a trawler owner joined Shri. Alemao in claiming that the trawlers from neighbouring Karwar were fishing in Goa waters and had gone away with huge haul. He warned that the agitated trawler owners would not allow the Fishery Minister to enter South Goa (Gomantak Times, July 2000). Shri. Jose Philip demanded a quick action from the Government. Dr. Wilfred D'Souza, the leader of the Nationalist Congress Party and MLA from Saligao and later the Deputy CM claimed that the Marine Fishing Regulation Act, 1980 was passed after a careful consideration of all the fishing aspects and it had gone to the Select Committee. He asserted that monsoon ban had existed for the last 5 years and none had protested against it, a fact which he said had to be brought to the notice of the High Court. Shri. Manohar Parrikar, MLA from Panjim wanted to know how the High Court had set the date and suggested an emergency amendment to provide solace to the mechanized fishing trade (Ibid).

The agitated members wanted to know why the courts have been intervening in the affairs of the State and whether the affairs of the state were run by Courts. "If that was the case, what was the role of the elected members"? they asked. The Chief Minister assured that the Government was committed to safeguard the interest of the fishing community and would look into matter and after consulting the State Advocate General would take the necessary action including an amendment to the law or even use the option of going for an appeal to the Supreme Court (Herald July 2000).

Thus, for the first time in many months the Opposition, the Government and the whole House were on the same wave length, as if they were stuck by a national calamity or disaster. It is beyond any body's comprehension how a ban of just 76 days would ruin and threaten the livelihood of the boat owners (who are well placed in the Goan economy). However, some MLAs like Jose Philip refused to accept that the trawler owners are well to do community. Mr. Jose Philip claimed that trawler owners are in huge debt, and have not been able to pay the loans. Moreover, they have to pay the hired outside labourer on daily basis as per the contract, whether there is work or not (Interview with Jose Philip 2000). As far as their argument about the protection of the interest of the fishing community, it is pertinent to ask whose interest anyway it was. Was it public or personal? The ecologists, environmentalists and scientists felt that the fish should be allowed to rest for 92 days when during the remaining period man ruthlessly kills to give way to his personal greed. Throughout the centuries man has been plundering nature's bounty without any remorse, all in the name of the Moloch of wealth.

The legislators in Goa should have seriously examined the High Court 'Interim Order', which underlined the long term gains – namely the protection of depleting marine resources. The High Court Order was based on the experts' recommendation but the legislators were not in a mood to accept neither the sanctity of the experts nor the wisdom of the learned judges. Many of the MLAs interviewed during the course of this study, though denied a judicial encroachment or confrontation but firmly asserted that business of the judiciary was to interpret the laws and not to make them and that the legislature was within its right to make, unmake and amend the laws.

The debate in the legislature was a precursor of the events that were to follow. The legislators were determined to act fast, against perhaps what appeared to them an odd judicial decision, two odd men against the rest of Goa. They had to come to the rescue of the fishing community.

MLA Jose Philip in an interview to the researchers took pains to explain that the experience of the sea does not require any scientific data and if one went by NIO data, fishing would not be possible any time in Goa. He opined that given the pattern of the weather and the early break of the monsoons, the High Court's ban up to 15th August was irrelevant and had to be rectified.

The High Court Judgment was given on 20th July 2000 and the amendment bill to the Marine Act was introduced on 21st July 2000. The Legislative Assembly of Goa granted leave to introduce the Private Members Bill piloted by the Congress member Churchill Alemao. This bill was titled: The Marine Fishing Regulation (Amendment) Bill, 2000 (Bill No. 31 of 2000).

It was published in the Official Gazette for general information in pursuance of the provisions of Rule 138 of the Rules of Procedure and Conduct of Business of the Legislative Assembly. The Bill was aimed to perpetually fix the prohibitory period for mechanized fishing from 1st June to 24th July every year, thus nullifying the High Court decision and secondly, proposed Amendment to certain sections of the Goa, Daman and Diu Marine Fishing Regulation Act, 1980.

For the benefit of the reader the bill is presented below (Ibid).

The Bill:

1. Short title and commencement :-

   (1) This act may be called the Goa, Daman and Diu Marine Fishing Regulation (Amendment) Act, 2000.

   (2) It shall come into force at once.

2. Amendment of section 4 :-

   (1) In the sub – section (2) of section 4 of the Goa, Daman and Diu Marine Fishing Regulation Act, 1980 (hereinafter referred to as principal Act) Clause (b) thereof shall be omitted; and

   (2) After sub-section (2) the following shall be inserted:-

   (3) With a view to conserving fish and regulating fishing on a scientific basis:-

   (a) Fishing by means of trawl nets, purse-seine nets, gill nets, etc. with mechanized fishing vessels and motorized fishing crafts, in the specified area along and on sea coast of the State of Goa declared as such by the Government of Goa by notifications issued earlier or that may be issued in future, shall be prohibited.

(b) Further, fishing by means of mechanized fishing vessels beyond the specified area declared by the Government of Goa as aforesaid, and up to the territorial water limits along the sea coast of the State of Goa shall be prohibited from 1st June to 24th July every year.

(c) Further, catching of juveniles of fishes such as mackerels, sardines, etc., in the specified area territorial waters along the sea coast of the State of Goa, as referred to above, shall be prohibited throughout the year.

Explanation: - "Mechanized fishing vessels" means a ship or boat fitted with mechanized means of propulsion which is engaged in sea fishing and includes a country craft and canoe fitted with inboard or outboard motor engaged in sea fishing.

3. Insertion of Section 27 A:- In the principal Act, after Section 27, the following section shall be inserted:-

"27 A Bar of jurisdiction of courts:- Notwithstanding any thing contained in any law, notification, order or decree of a court or any other authority and not withstanding the proceedings in any court, tribunal or any other authority shall have Jurisdiction in respect of matters contained in sub-section (3) of section 4 of this Act".

The Bill in its statement of objects and reasons stated that the Government of Goa had issued a notification dated 8th April, 1990 to achieve the objectives of section 4 of the Goa, Daman and Diu Marine Fishing Regulation Act, 1980 by which a ban on mechanized fishing beyond the specified area and within the territorial waters along the sea coasts of Goa was imposed from 1st June to 24th July every year. The bill states that the ban period was kept at 54 days so as to uphold the interests of different sections engaged in fishing. The bill further stated as the Government notification was under question in a Court of Law and as the public at large in the State of Goa were facing serious hardship, it had decided to affirm the rights of the people of

Goa through their elected representatives so that the ban is kept for only 54 days through legislation in place of Government notification. As the bill had no financial implication, no financial memorandum was appended (Ibid).

While moving the Bill Alemao argued that extension of the ban beyond 24th July would be grave injustice to the trawler owners, many of whom were struggling to repay loans. Secondly, he said the solar variety of prawns and ribbon fish were available only in July and there was no guarantee that these species could be harvested in abundance later. Reacting to the NIO report Churchill Alemao argued that if one goes by scientific information provided by the institute and if one goes by the period mentioned in the study of marine species breeding season, one would get the impression that fishing was to be totally banned. He questioned why the ban should be extended beyond 24th July when in a place like Nagapatnam there was no ban at all and when trawlers from neighbouring Karwar were fishing on the Goa coast. Honorable Member must have put his tongue in the cheek when his colleague Francis Silveria's trawler, a few days later, was impounded by the Fisheries Department of Maharashtra for the violation of a ban on mechanized fishing imposed by the State until 15th August. Churchill Alemao pleaded with the House that the bill should be urgently passed as boats from the neighbouring States were quietly fishing in the Goa waters so also when the foreign fishing vessels were allowed under Central law to fish on the Goa coast. He warned that the solar prawns which had come with the wind to Goa waters would not remain long. He reminded the House that when the trawlers used to fish by 15th July, they used to get 15 to 20 or 30 tonnes of prawns (Unpublished, Official Legislative Debates July 2000).

Participating in the debate Chief Minister Sardinha urged Alemao to appeal to his counterparts in the fishing trade not to catch the juvenile of the fish and use the right size nets. Shri. Sardinha admitted that fish was vital to the Goans and fishing was necessary if the people were to get fish in the market. But quoting Gandhi he said, "There is enough in this world for everybody's need but not for everybody's greed." So Government would not hesitate to enact another Bill to protect the juvenile of every species in the sea and river (Ibid).

Shri. Sanjay Bandekar MLA from Canacona suggested that there was no need of another legislation as the provision of the Central Marine Regulation Act had given power to the Government to take stringent measures against such erring fisher folks. (The Navhind Times July 2000).

Very rarely do the Private Member's Bills sail smoothly in the ocean of legislative debates and amendments. But this Bill invited very little debate or no debate at all. There was no dissenting voice. Being a minor amendment to the Goa, Daman and Diu Marine Fishing Regulation Act, 1980, (The Act was passed after great deliberation and referred to a Select Committee); it was taken for immediate consideration. The mover of the Bill and just two members spoke. Then the Speaker submitted the Bill to the House for the clause by clause vote. Finally, Shri. Alemao moved that bill no. 31 of 2000 – The Goa Daman and Diu Marine Fishing Regulation (Amendment) Bill, 2000 be passed. The motion was put to vote and passed unanimously by Voice – Vote.

The urgency of passing the Bill was seen by the fact that the bill was not in the agenda of the business of the House. The Rules of Procedure and Conduct of Business of the House require prior notice of the bills that are to be taken and they must be listed in the agenda of the day. Moreover, there is special procedure of selecting private member's bill. In this case Rule No. 135 dealing with precedence of Private Member's Bill and 141 regarding advance circular were wavered as per the wishes of the House, which is supreme and over which no authority can sit in judgment. The Bill though passed by the Legislative Assembly on 21st July 2000, generated a lot of public resentment as some healthy conventions were set aside. The Legislative Assembly in Goa created a history of a kind by passing an Amendment Bill in such hurry and that, too, on an issue on which the Panjim Bench of the Bombay High Court had just given an Interim Order. Generally, issue of this type give a lot of mileage to the Opposition, which normally calls for a censure motion or a walkout or even the resignation of the Government. But what was strange about this Bill was it came from the Opposition Benches, it was allowed by the Speaker and unanimously voted by both the ruling and opposition.

The hastily passed amendment was claimed to be in public interest. In the course of this study, a majority of the MLAs that were interviewed justified the wavering of rules and the passing of the bill. They claimed that they had acted in a larger interest. They had passed the legislation, as the particular type of prawns were available only in this particular season and the catch would help the trawler owners to pay the heavy bank loans, failure of payment would mean bank seizure of the vessels. One MLA, however, openly admitted that he was against the bill but as his party had issued a whip he was bound by it. Many members of the legislative assembly mentioned that in the absence of a uniform ban all over India, they saw no reason but to support the bill. (Interview with MLAs, July 2000).

The legislators passionately pleaded that they had to protect the interest of different sections of the society but the Bill far from being welcomed, it only generated public fury. The clause 3.27A, which debarred the Court from exercising its jurisdiction in respect of matters, contained in sub-section (3) of section 4 of the Act relating to mechanized fishing vessels in the specified area and beyond the specified area proved to be most controversial. The amendment had fixed the monsoon ban period from 1 June to 24 July every year in defiance of the High Court Order and protected it from any future judicial review. Though the MLAs that were interviewed during the course of this study denied a conflict between the Judiciary and the Legislature, some members expressed their displeasure on what they called encroachment on the turf of the legislature (Gomantak times July 2000). Weather or not the two wings of the Government were on war path, the clause that debarred judicial review of the monsoon ban was a reminder to the judiciary that it was free to give its own interpretation but the legislature was supreme to modify any law that it thought necessary in what it felt public interest. However, one MLA openly told the researcher that he was happy with the decision of the High Court but had to vote for the bill under the instruction of the party whip (Interview July 2001).

The Amendment Bill reopened the conflict between the mechanized Boat Owners and the Raponkars, the traditional fishermen. The all Goa Fishermen Union – Geonchea Raponkaracho Ekvott (GRE) termed the bill as 'scar in the history of Legislative Assembly as it has been passed by violating all rules and procedures. The GRE accused Alemao, who is a trawler owner, of misusing his position as a legislator, to foster his personal business interest against the public good. The General Secretary of GRE,

Mathany Saldanha stated that the "conservation of marine resources is far more important to protect the fishing communities than opting to destroy the roe bearing fish for profit motive in the name of fishing community (Mathany Saldanha Interview July 2001). He further argued that the Marine Regulation Act was introduced not only to protect the traditional fishermen but also to conserve the depleting marine resources due to over fishing. He added that High Court decision had the endorsement of the traditional fishermen for whom fishing was their only livelihood and therefore, GRE wanted the "Conservation of natural resources to be a priority, above vested interests and short term games of profit motive".(Herald 25th July 2000). He warned that under no circumstances GRE would accept the "unconstitutional amendment brought in by the Legislative Assembly and would call upon the Government and the President of India to inform them of the "Unconstitutional activities of the Goa Government" (Ibid).

In an interview to these researchers, Saldanha accused the Goa Government of destroying Marine ecology as it had failed to implement 5 kms ban and also the monsoon ban. He questioned how people could trust corrupt legislators who had betrayed the peoples' interest and trust reposed in them. Indeed legislators are there to serve the long term interest of the people and not the vested interest of the legislators themselves. The action of the Goa legislators strengthens the credibility of Rousseau's distrust of the representatives and his condemnation of Representative Government. Sometimes, representatives legislate not to promote public good but their own interest. It also confirms a famous observation of Lord Bryce that the frequent rejection by people of measures passed by the Assembly shows that latter does not always know or give effect to what has proved to be the real will of the people. Critics of Representative Government hold that power and money are the preoccupation of the representatives and the interest of the representatives and those represented are not always identical.

The press too was furious. In an editorial of 22nd July 2000, the Gomantak Times, English daily, charged that fishing lobby was trying to fish all the year round and unable to do so, they managed to reduce the period progressively to just 54 days. The Editorial further stated that fishing was a traditional occupation. It was halted during monsoons only to be resumed after Narali Pournima, which came in the Shrawan month. The spawning season of the fish coincided with the monsoon on the west coast. Just as people in the ancient past kept the land fallow once every year to replenish its fertility, in the same way fishermen took break at fishing, so that the fish could spawn. Fishing activities resumed after Narali Pournima with breaking of coconut to the sea God. The editorial concluded that the Interim Order was in keeping with the wisdom of the fore fathers. Many of the MLAs interviewed during the course of the research dismissed the Narali Pournima as more of a cultural taboo. One MLA even questioned the wisdom of following the Hindu calendar where for every other matter the English calendar was observed (Parrikar in an interview, 2001). Undaunted by the public reaction, Government was determined to make the bill into law. It was sent to the Governor for his signature. Meanwhile, gratified Trawler Owners Associations of Cutbona, Vasco and Tiswadi thanked the then Chief Minister Sardinha for passing the Bill. Gomantak Times reported that 50 bags of prawns were distributed to each of the 40 MLAs for their support. When these researchers asked the MLAs to comment upon this allegation, some denied, others admitted it, others scoffed at the charge saying that the 50 bags of prawns was too small a bribe for a piece of legislation passed in public interest.

But the saving Grace came when Governor M.Fazal refused to give his consent to the Amendment Bill. He found the clause in the bill barring the jurisdiction of the court vis-à-vis the ban period objectionable.

But the crisis was far from over. The boat owners were confused, on one side there was the Legislative Amendment, on the other side, there was the High Court Order. Perhaps, believing in the legislative supremacy, some of the boat owners sent their boats to the sea. Government on its part showed little desire in implementing the court order or even patrolling the sea. On 31st July 2001, the local newspapers brought photos of trawlers held captive by traditional fishermen at Agassaim. The representatives of the Ramponkars in a complaint to the police stated that about 130 trawlers were fishing in the river Zuari on 30th and 31st July in defiance of the court order and they badly mauled their stake nets causing irreparable damage.

Government's failure to protect the law angered the public. Advocate Norma Alvares in a petition to the Bombay High Court informed the court of the violation of 20th July Court Order. She presented before the Court pictures of trawlers going to the sea. She further appealed to the Court to seal the jetties in the State. The Court turned down the demand for sealing the jetties but ordered the immediate suspension of fishing licenses of all trawler owners in the State till 15th August 2000 and barred seven jetties from being used by the trawlers for fishing purposes (Ibid August 2000).

Thus, the High Court once again showed itself as the protector and upholder of the interest and welfare of the people. The Government should have upheld the order of the High Court and should have taken effective measures to implement it but it turned a Nelson eye. The conflict between the two wings of the Government encouraged the influential and powerful boat owners to fish in the troubled waters. The Government apathy to the Court Order only intensified the old conflict between the Ramponkars and the Mechanized boat owners. The Government Order to reduce the prohibitory period and the subsequent Legislative enactment to nullify the High Court Order only proves that the Government in Goa had no commitment to conserve marine resources which are day by day depleting, nor it had public interest at heart but it identified itself with specified interest. The Goa Government failed to be the arbiter or mediator in the fishing dispute but made itself a party to the dispute. This is not good governance.

The Fisheries Department has been vigorously pleading that 54 days monsoon ban is adequate enough to protect marine life. In a supplementary affidavit filed before the High Court on 16th August 2000 (Goa Foundation, Fishing File), the Fisheries Department maintained that as per the scientific data on the spawning period of the fishes of the sea coast of Goa, submitted by the Central Marine Fisheries Research Institute attached to Indian Council of Agriculture and Research reveals that the monsoon ban period gives adequate protection to the major species of fishes like Indian Mackerels, Indian Sardines, Ribbon fish, etc. It further stated that the ban not only gives protection from the ecological point of view but from the point of socio-economic needs of the local population, which is dependent on fish as the cheap protein food at affordable prices.

In another affidavit filed on 29th August 2000, the Fisheries Department argued that in the light of the High Court Order August

2000, (vacating its earlier interim order of 20th July 2000) the Government and the Cabinet on the basis of factual and scientific data, at its meeting of 25th August 2000 took a policy decision that fishing period since 1995 has been beyond 24th July (Herald 9th April 2001). Herald, the popular English daily blasted the Government's decision. It charged that Parrikar Government had succumbed to the trawler owners' powerful antics and no matter which party or coalition was in power, the influential trawler lobby headed by Benaulim politician continue to enjoy the clout. The paper charged that a delegation of trawler Owners met the Chief Minister Parrikar and lobbied hard for reducing the ban period. On 1st June 2001, Government announced a complete ban during monsoon with effect from 1st June for fishing in the territorial waters of Goa by means of trawlers, purse-seine, and gill nets. Fishermen were also informed that Government of India had also imposed a ban on the operation of all deep sea fishing vessels in the exclusive economic zone (EEZ) of Goa Coast from June 1st to July 24th, 2001 (Ibid June 2001).

Meanwhile Advocate Mrs. Norma Alvares through a letter 6-6-2001, informed the Under Secretary, Government of India, Ministry for Agriculture about the PIL Writ Petition pending before the High Court (Panjim Bench) challenging the decision of Goa Government to reduce the ban on mechanized fishing to 24th July, from the earlier 31st August, under the pressure from the trawler owners. She informed the Government of India that the High Court after obtaining information on the breeding season from the NIO and the Central Marine Fisheries Research Institute, issued an Interim Order banning mechanized fishing from 1st June to 15th August. The Goa Government in its reply had conceded that economic consideration influenced its decision and that it had taken a policy decision to impose the ban on fishing from 1st June to 31st July. The High Court not satisfied with the decision extended the ban up to 15th August and postponed the matter for final hearing. Mrs. Alvares was drawing the attention of the Government of India to the fact that mechanized fishing ban period was up to 15th August and not to 24th July as per earlier Government Notification.

With the matter still pending before the High Court, the bitterness between Goenchea Ramponkarancho Ekvott (GRE) and the Trawler Owners is ever-growing. The General Secretary of GRE,

Mathany Saldanha in an interview to an English Daily the Navhind Times, 17th July 2001 complained that the traditional fishermen were marginalized and the mechanized boats were catching the fish in the coastal waters within 200 metres.

The matter is still pending before the High Court but the issue has invited a nation wide interest. Scientists, academicians are debating over the feasibility of adopting a National Fishing Policy.

CHAPTER VIII

MANAGEMENT OF FISHERIES

Globally, fisheries are recognized as a vital source of food, employment, trade and economic growth. It has been reported that global production of marine fish is decreasing. The Annual rate of increase in marine catches decreased to almost zero indicating that, on an average, the world's oceans have reached their maximum production under the present fishing regimes. The FAO data shows that globally fishing fleets have increased rapidly between 1950s and 1990s and so also the adoption of more advanced technology. During the last few years, the number of fishing vessels have decreased in developed countries but have increased in some developing countries (Sinclair and Valdimarssim: 2003). This makes it obligatory on the part of the developing countries to work out a sound fishery management policy.

The Fisheries Management has a long history. Till about the end of the 19th century, it was widely believed that the living resources both in the oceans and seas are limitless. At that time, the fishermen communities knew well the management of their fisheries. Many had also developed a system for managing fisheries. However, these regimes of fisheries were known only to them as they were "recorded only in their oral and behavioural traditions" (James R, Mchoodwin, 1990:66).

However, the Industrial Revolution and subsequent industrial growth began to threaten the fisheries by polluting them. The fisher folk communities' world over concerned about these developments began to pressurise their respective governments to protect their fisheries as they had no control over these developments. The Technological Revolution in course of time brought more advanced and effective techniques into the fisheries. The change from sail to steam and then to gasoline and diesel engines gave increased power and ability to trawlers to move into deep fishing. The fishermen were able to fish in ocean regions never fished before. They could remain in the sea for many days. Thus, the industrialized fishers began to catch fish on large scale. This was necessitated by the rapidly growing population and the urge to earn more foreign exchange. But by the end of the 19th century, the negative effects of these developments in the fisheries sector began to appear. In many regions, particularly, in the North Atlantic and North Sea Fisheries, certain marine resources began to deplete. Despite increased catch in the initial years, the fisheries sector confronted with several common problems like overexploitation, environmental degradation, fishing conflict, etc. The widespread belief that the marine resources of the sea are inexhaustible began to crumble down. The end result of all this was the "birth of modern fisheries science and the modern practice of fisheries management" (Ibid 1990:67).

Today, the fisheries management has not remained just a local, regional or national issue but it has become an international concern. Almost all countries of the world having sufficient fisheries have realized the urgent need of having better management of their fisheries.

A sound fishing management policy, therefore, involves fishery development, the regulation of the mesh size, the system of licensing and other related regimes. Fishery development implies the expansion of fishery effort without affecting the sustainability of the fishing activity. It may be defined more broadly to include, in addition to the expansion of fishing effort, the improvement in post-harvest technology, marketing and transportation of fishery products as well as the provision of infrastructure and other related facilities (FAO: 1982).

The indirect control of effort includes the case in which a management authority does not get involved in the control of effort but simply creates the appropriate environment for its control by the fishermen themselves (e.g. community property rights).

The overexploitation of fisheries, sooner or later, will lead to the regulation of fishing effort. This will lead to surplus labour and capital. A sound fishing management policy will have to look into this matter.

If the policy objective is to maximize the economic benefit to the national economy from the fishery, then we have to achieve the optimum rate of exploitation. The optimum rate of exploitation is defined by the Maximum Economic Yield (MEY), that is, the maximum sustainable surplus of revenues over fishing costs. Alternatively, MEY may be thought of as a modification of Maximum Sustainable Yield (MSY) to take into account the value of the fish caught and the cost of catching it. The fishery is said to be underexploited in the economic sense and requires further development if the actual catch falls short of MEY due to insufficient effort. The fishery is said to be overexploited in the economic sense and calls for management if the actual catch falls short of MEY due to excess fishing effort (Panayotoy:1982).

Competitive exploitation of common property fish stocks is wasteful because each fisherman or company director does not take into account how his fishing effort affects the size of fish stocks and thereby their growth and the unit cost of fish. Hannesson (1993) gives two ways of dealing with this, which are as follows:

1. Establishing property rights to fish stocks. Under a system of property rights, no one would have the right to fish from a stock except the owner and those who are explicitly authorized by the owner to do so. The owner could be a person, a company, or a collective of individuals. Clearly, it would be in the owner's best interest to take into account how fishing affects the growth of the stock and the unit cost of fishing.

2. Control of the fish stocks by the public authorities. Clearly, the access to the fish stocks must be limited in some way in order to avoid the over exploitation resulting from open access. In this type of solution the fish stocks would remain common property in the sense that no one could claim exclusive rights to them unless being granted such rights by the public authorities.

PROPERTY RIGHTS TO FISH STOCKS

The owner of a fish stock has a strong incentive to limit fishing to whatever level to maximizes profit, whether he cashes in the profit by fishing himself or by selling or leasing to others the right to do so. The profit-maximizing fishing effort will also be the socially optimal effort, provided the prices on the basis of which the profit is being maximized satisfy the following conditions:

1. The owner of the fish stock must not be able to influence the price of fish or the price of inputs but must take them as being given by the market.
2. The prices of fish and fishing inputs must reflect the social opportunity cost of fish and inputs.
3. The discount rate used by the sole owner must be equal to the social discount rate.

However, Hannesson (1993) gives reasons why the property rights solution has seldom if ever been tried and is not likely to be tried on an extensive scale. Most fish migrate widely, and the possibility of 'fencing them in' and effectively claiming ownership is small. For fish that straddle international boundaries, the property rights solution would run into difficulties, as all private ownership depends ultimately on the legal apparatus and enforcement of a sovereign state. For fish that stay within the economic zone of one single state, the property rights solution is likely to be impractical and undesirable except for the most stationary stocks. For migratory stocks, the property rights would have to be defined over a wide area in order to provide an effective control over each stock. This might give the property rights holder an undesirable degree of control over the market for fish or fishing inputs.

There are two ways in which property rights to fish stocks could be implemented. One is to make it illegal for anyone but the owner to land fish from a certain stock. This requires that it be possible to identify from which stock the fish originates, either by genetic characteristics or by knowing where it was taken. This is often possible. The other way is by territorial rights. To confer full control over a stock, the territorial right would have to apply over the entire area which a stock inhabits. Since fish stocks vary greatly in their migratory habits, an array of territorial rights differentiated according to stocks would have to emerge in order to ensure full control over all stocks.

It is possible, however, that territorial use rights in fisheries (TURFS), defined over some territory without any reference to the migratory habits of fish, would facilitate an efficient use of fish stocks. Even if such rights would not give any single holder of a territorial right full control over any single stock, they would provide an incentive to conclude mutually beneficial agreements on how to harvest the stocks. Collectively the holders of territorial rights would be interested in maximizing the aggregate profit of the fishery. Territorial rights would provide an incentive to do so, as they would limit the number of parties with access rights.

In Japan, fishermen's cooperatives have been given exclusive rights to inshore resources. These rights come close to amounting to ownership of the resources. The fishing rights appear successful in promoting efficient utilization of the resources. The fishermen's cooperatives regulate their members' access to the resources (others are excluded by the rights) and engage in stock enhancement such as the building of fish shelters and artificial reefs and the operation of hatcheries. There would be little incentive to engage in such activities without an arrangement which ensures that those who plant have the right to harvest (On the Japanese fishing rights, see Asada et al. (1983), Ruddle (1984), Yamamoto and Shot (1972).

Next alternative is government regulation for fishery management. It has been tried and found that it often fails.

WHY GOVERNMENT REGULATION OFTEN FAILS?

Most regulations of fisheries have been initiated when free access has caused severe and obvious depletion of the resource base, leading to low incomes or unemployment in the fishing industry. At this point there has emerged, at least, a degree of consensus that the resources must be safeguarded and replenished, in order to raise incomes and maintain employment in the industry. Fisheries biologists have been called upon to give advice on how to manage the resources to attain this goal. The regulations imposed to do so have typically aimed at limiting the catch of fish in order to protect the fish stock, such as closing the fishery for a certain period, prohibiting the use of certain kinds of fishing gear, etc. which has been tried in Canada, Norway, Indonesia, etc.

The challenge lies in devising a regulatory mechanism that emulates a perfect market and is as independent as possible of political decision-making. One step in this direction would be to set up a public management authority with an explicit mandate to maximize the rents in the industry and to reward its managers according to how well or badly they perform. An agency like that would most likely see an advantage in relying on efficiency-promoting mechanisms like individual transferable quotas.

Since moving towards increased efficiency from a position of free access and no rent implies some reduction in fishing effort, this will be difficult to attain with the consent of the industry, unless those who leave the industry get a compensation making them no worse off than they would be, if they had continued fishing.

Irrespective of whether fishermen deserve to be compensated for the losses they would incur for rebuilding a fish stock (after all, they depleted it in the first place), it is important to note that such compensation is possible, and the very fact that it is possible is itself the reason that a replenishment of the stock is indeed worth while. Increased economic efficiency means that a surplus will emerge, in this particular case an excess of revenues over costs. Those who participate in the fishery at the time when the replenishment of the stock starts can share in this gain, so that no one need come out worse than if the transition to the more efficient situation was never made. Since the gains take time to emerge, the government or the industry itself would have to advance funds to compensate potential losers. A practical example of this mechanism is the introduction of individual transferable quotas in the inshore fishery in New Zealand. About 20 per cent of the fishing quotas allocated on the basis of catches in the immediately preceding years was bought back from the industry in order to rebuild the fish stocks by reducing the total catch.

NEED FOR MANAGEMENT OF FISHERIES

Human concern for a proper fisheries management has resulted in range of initiatives at the global, regional and national levels. Overfishing and excess fleet have been identified as worldwide problems that call for socially acceptable and effective solutions. At the national level, every country is responsible for proper conservation, management and development of its national fishery. But, as in India, fishery is a state subject it is the responsibility of each state to work out its own Fishery Management Policy.

The conflict in Goa between the traditional and mechanized fishing communities continues over space, fishing ground and fishing regimes. The artisan fishermen accuse the mechanized boat owners of intruding in their "specified area" and even fishing within 200 meters from the shore (Mathany Saldanha NT 17-6-01), an accusation vehemently contested by the mechanized boat owners. The artisan fishermen to prove their point on many occasions impounded mechanized boats in the fishing villages of Agasaim and Colva. They charge that the mechanized boats strayed into their waters damaging their gear and nets. Of the 99 artisan fishermen interviewed in the process of this study, 70 per cent of them acknowledged that such incidents have taken place and no compensation was paid to them for the losses incurred by them. They also charged that the mechanized boat owners were only interested in making profit and have no concern for ecology or environment. Ignoring the monsoon ban they caught fish, which was in the process of spawning, or their juvenile, thereby causing depletion of the fish in the short term and in the long run threatening the very sustainability of the fish resource itself.

The onsets of monsoons every year causes a great concern and anxiety among the artisan fishermen, ecologists, environmentalists and the public at large whether the old conflict will be renewed between the two major contenders in the sea.

The Government as an arbiter has to transcend above class and interest. It has to see that both the communities conduct their activity in an atmosphere of peace and harmony and that all the contenders observe the rules of the game and each one gets his share of the fish resources. It is also the responsibility of the Government to promote sustainable fishing resources and has to make available the improved technologies which could help in identifying new fishing zones so that every body get something and yet there will be plenty for the future.

These issues call for a sound Fisheries Management Policy. It implies the rational exploitation of fisheries. It involves allocation of fish resources, formulation of rules and their enforcement. In short there must be effective fisheries management based on the notion of sustainability, efficiency and equity, through effective Monitoring Control Surveillance regimes. In the absence of the sound fisheries management there will be the "tragedy of the Commons".

As per the FAO definition "Monitoring" involves collection of data on biological, economic and social aspect of fisheries and basic information of fisheries, vessels and gears and nets used (Workshop on Monitoring, Control Surveillance – Goa India 12-17 Feb. 2001). This will help to know the changes in catch rate and fish sizes. This will also help to keep a watch on fishing trends and patterns, which will provide input into fisheries. In this regard, Government should monitor the fishing communities and their problems.

Fisheries Management Control is necessary so that resources are available for the future benefit of all resource users. Management Controls are applied through a variety of management instruments in order to regulate the fishery. Surveillance is another important component of fisheries management. It relates to policing and supervision of fisheries activity. There is a growing consensus over the need to strengthen patrolling by the Government of the fishing activity and the enactment of more vigorous regimes. "Fishery Department and the enforcement officers feel the need of Vessel Monitoring System". The conventional MCS involves a lot of expenditure, and basically it is naval aerial surveillance. The VMS is not an alternative to conventional but it would strengthen the conventional monitoring and control system. One of the important instruments is through legislation. The Governments passed the Goa, Daman and Diu Marine Fishing Regulation Act in 1980. The main provisions of the Act relating to fishery management are as follows:

1. ***Power to regulate the mesh size of fishing nets-*** The Government may regulate fishing on a scientific basis by an order notified in the official Gazette and regulate or restrict the size of mesh of fishing net. The recommended mesh size for prawns is 20 mm. and 22 mm. for other species of fish. This is needed to conserve fish resource.

2. ***Licensing of fishing nets-*** The owner of fishing net is required to make an application to the authorized officer for the grant of license for using fishing net.

3. ***Licence period***-A license granted shall be valid for the period specified therein or for some extended periods as the authorized officer may think fit to allow in any case.

4. ***Prohibition of fishing using fishing vessels, which are not licensed***-No person shall, carry on fishing in any specified area using a fishing vessel which is not licensed under section 6, provided that nothing in this section shall apply to any fishing vessel, which was being used immediately before the commencement of this Act, for such period as may be specified by the Government by notification in the Official Gazette.

5. ***Control on manufacture, sale or use of fishing nets***- No person shall after the commencement of the Act manufactures or uses for fishing any net in contravention of any order passed under section 7 or without obtaining any licence as required under section 8.

6. ***Cancellation, suspension and amendment of licenses***-

(a) If the authorized officer is satisfied either on a reference made to him in this behalf or otherwise that- that a licence granted under section 6 or section 8 has been obtained by misrepresentation as to an essential fact; or

(b) The holder of the licence has, without reasonable cause, failed to comply with the conditions subject to which the licence has been granted or has contravened any of the provisions of this Act, or any order or rule made there under, then, without prejudice to any other penalty to which the holder of the licence may be liable under this Act, the authorized officer may, after giving the holder of the licence a reasonable opportunity of showing cause, can cancel or suspend the licence or forfeit the whole or any part of the security, if any furnished for the due performance of the conditions subject to which the licence has been granted; subject to any rules that may be made in this behalf, the authorized officer may also vary or amend a licence granted under section 6 or section 8.

PENALTIES

1. ***Power to enter and search fishing vessels-*** The authorized officer may, if he has reason to believe that any fishing vessel is being or has been used in contravention of any of the provisions of this Act, or of any order or rule made there under or any of the conditions of the licence, enter and search such vessels and impound the same and seize any fish found in it.

2. ***Disposal of seized fish, etc.-***

   (1) The authorized officer shall keep the fishing vessel impounded under section 18 in such place and in such manner as may be prescribed.

   (2) In the absence of suitable facilities for the storage of fish seized, the authorized officer may if he is of opinion that the disposal of such fish is necessary, dispose of such fish and deposit the proceeds thereof in the prescribed manner in the office of the adjudicating officer.

3. ***Penalty-*** The adjudicating officer shall, after the enquiry under section 20, decide whether any person has used or caused or allowed to be used any fishing vessel in contravention of any of the provisions of this Act or of any order or rule made there under or any of the conditions of licence and any such person on being found guilty by the adjudicating officer, shall be liable to such penalty not exceeding:-

   (a) five thousand rupees, if the value of the fish involved is one thousand rupees or less: or

   (b) five times the value of the fish, if the value of the fish involved is more than one thousand rupees:

   (c) five thousand rupees, in any other case, being a case not involving in any fish, as may be adjudged by the adjudicating officer.

Several Rules and Regulations have been issued on the basis of the marine fisheries act. For example, the Goa (brackish water)

Fish Farming, 1991, and Forest and Agriculture department related orders. However, some communities have expressed their reservation over the Marine Fisheries Act of 1991. It is submitted that this legislation could be reviewed so as to ensure the protection of all the interests involved, the consumers as well as the resource providers.

BAN, CONSERVATION AND SUSTAINABILITY

The monsoon-fishing ban in Goa has generated a nationwide debate. The trawler owners charge the fishing vessels from neighbouring states that they fish in Goa's waters during the monsoon ban. The other fishing states charged that Goan fisherman fish during monsoon ban in their waters. These charges and counter charges have made some people propose a uniform fishing ban all over India. A proposal much contested by experts and marine scientists. As the monsoon induced nutritional enrichment in the sea varies from coast to coast, the ban cannot be uniform to both the coasts. However, it is essential to have a uniform ban period, separately for each coast during monsoon period so that the gravid fishes are not caught during their peak-breeding season that would help in sustenance period.

Another factor, which is responsible for nutritional enrichment of coastal waters is upwelling, which also coincides with monsoon period in many places.

Many pelagic fishes are migratory in nature and have a prolonged breeding period spread over the year with one or two peaks coinciding with monsoon.

Bottom trawling during ban period should be prohibited as it can wipe out the resident demersal fish population.

During monsoon and post – monsoon period, majority of the fish breed. While in the west coast, monsoon period extends from May end to September, the east coast receives monsoon rains during October to December. Any sound fishing policy should reflect this reality.

In Kerala and North-Eastern states both South West and North East monsoons are active. The river brings in a lot of nutrients to the sea during monsoon and in turn this turbid water is congenial for the production of plankton, which forms the main food of young ones. This is the reason why a ban period is necessary during the monsoons when shoals of mature fishes come to inshore waters for breeding so that the young ones do not feel the shortage of food.

Tony Pitcher (1998) suggests the need of reinventing fisheries management to overcome fisheries crises. He suggests following reforms that could ensure sustainable, well-managed and ecologically sound fisheries.

1. Strengthen national, regional and international capacity to manage the marine fishes. Government must allocate sufficient funds to develop the institutional capacity. The fishery management at all levels must be relived from political interference.
2. Focus management programmes on limiting effort and restricting the access to fisheries.
3. Enact and implement recovery plans for depleted species.
4. Reduce and eliminate the subsidies that sustain commercial fisheries.
5. Accelerate programmes for decommissioning excess fishing fleet capacity.
6. Expand programmes for retraining fishers displaced by overfishing.
7. Develop social and economic incentive for sustenance well-managed fisheries.
8. Reduce the footprint of northern fleets on fisheries in developing countries.
9. Eliminate the destructive fishing practices such as the use of poisons and explosives.
10. Reduce and eliminate by-catch of marine wildlife in commercial fisheries.

Goa, though one of the prominent fishing states of the west cost, is yet to devise a sound fishery management policy which needs the twin challenges of sustainability and development. The fishery development is still at the primary level. The Government initiated fishery research is at subsistence level. The environmentalist and ecologist have expressed a great concern over the depletion of fish stock and the increase in fish famines. The Goa Government would have to do a thorough research on this issue and then work out its response and draw a recovery plan.

It has to work out a more effective monitoring system. The Directorate of Fisheries has been complaining that it is not adequately staffed or it has no necessary equipments for monitoring and surveillance. It has just one patrolling boat for supervision and enforcement of fishing rules and regimes. The Government will have to increase the patrolling facilities so that all the contenders in the sea keep to the rules of the game and therefore, all types of social frictions are avoided. The Government should also take measures to increase the productivity of stocks and the reduction of the wasteful use of scarce resources.

CHAPTER IX

CONCLUSION

The over fishing of the world's major marine fisheries was thinly disguised for many years by the overall figures for the global marine fisheries catch, which increased from about 20 million tons in 1950 to 82 million tons in 1989 (Porter:1998). These figures did not reflect the serious decline of the high-value species—those most preferred by the consumers and bringing the highest prices. The reductions in catch of these popular fish in 1992 compared to the catch levels in their peak year were extremely marked: Atlantic cod fell by 69 per cent , Cape hake by 82 per cent ,haddock by 80 per cent, silver hake by 88 per cent , greater yellow croaker by 80 per cent, and Atlantic herring by 63 per cent (Weber: 1994). Since 1983, increases in global catch have come primarily from five low-value species that represented only about six per cent of the catch by value (FAO 1995). By the beginning of the 1990s, about 70 per cent of the world's conventional fish species were either fully or heavily exploited (44 per cent), over exploited (16 per cent), depleted (five per cent) or slowly rebuilding after having been depleted earlier (three per cent) (FAO:Ibid)

Similar is the case with the Indian fisheries sector. The marine fisheries sector in India has witnessed a phenomenal growth during the last five decades both quantitatively and qualitatively. The subsistence fishery during the early 50's produced about 0.5 million tonnes annually and increased to around 2.7 million tonnes by 2003 (Srinath: 2003). This increase is the result of improvements in the harvesting methods, increase in the fishing effort and extension of fishing into relatively deeper regions. The increased effort over time and space is the consequence of ever increasing demand for marine food both from external and internal markets. This phenomenal growth also brought in imbalances in the exploitation across the regions and among the resources. Besides, with production levels for most of the commercially important resources showing signs of approaching saturation levels, inter sectoral conflicts increased due to competition to exploit the common resource. Fleet size and operations underwent quantitative and qualitative change. Traditional boats are being increasingly motorized and the mechanized sector operating with trawlers and gill-netters are resorting to multi-day fishing, thus contributing to increased fishing pressure (Srinath: 2003).

In tune with the trends at the national and international level in fisheries, the fishery sector in Goa also witnessed rapid growth and changes after its liberation. From the traditional fishing the state has gone to mechanized fishing- trawling and purseining. In 1960-61, there were only four mechanized boats and 4125 country crafts. But in 2004, the number of mechanized boats increased to 1134. The Government of Goa promoted mechanized fishing by giving them subsidies and other facilities.

These mechanized vessels like trawlers and purse seines drastically affected the traditional fishermen's activity in Goa. The study revealed that mechanization was profitable in the initial years. The fishermen used to make three trips every day and within an hour their trawlers used to be full. But today even after long sailing, sometimes, they do not get enough quantity of fish. Some fishermen blame the government for encouraging them to adopt mechanization. Some of them blame themselves for over fishing by way of making 3-4 trips a day. Majority of the respondents feel that deep-sea fishing is responsible for reduction in the fish stock in the sea. The wrong priorities and short vision of the policy makers and planners were responsible for these fall outs of fisheries development.

It is found that though the government has stopped giving new licenses for mechanized boats, the construction of new boats is not banned. Majority of the respondents said that in place of the old crafts new boats are replaced and the government does not object this. Besides, new boats are being built and operated without the feasibility report. They get registered after 2-3 years by bribing the concerned authorities and the political leaders. It is noticed that some trawler owners do not belong to the fishermen community. Soon after mechanization of fishing, many entrepreneurs from non-fishermen community ventured into this business to make lucrative profits. The object of such entrepreneurs is to make maximum profit at any cost. They lack the regard and knowledge of the sea, and its resources, which the traditional fishermen had. They were aware of the need to preserve the fish for future generations. For this, they had adopted the traditional methods. The new entrepreneurs motivated by high returns for their huge investments, are engaged in the ruthless exploitation of marine resources.

The lack of political will and determination to take and implement corrective steps sincerely and efficiently has further aggravated the situation. Many political leaders own the trawlers as they have also entered into this modernized fishing occupation. Therefore, they become the stumbling block in the way of taking measures for sustainable fisheries development. Due to these and many other factors the fisheries managers and administrators have utterly failed to manage the change within the fisheries sector for sustainable growth. The development of fisheries characterized by conservation, sustainability and freedom from pollution requires collective effort of all the stakeholders in the fisheries sector, which is lacking to a great extent.

The government resorted to modernization process of the fisheries sector without creating level playing ground for the artisan fisher folk. They did not possess the pre-requisite qualities of modernization like capital, education, basic skills, etc. Therefore, the modernization process initiated and supported by the government had no meaning to the artisan fisher folk. They could not take the advantage of these processes. This resulted in the entry of non-fishermen in this sector whose sole motive was earning maximum money. Having no sensitivity to the issues like sustainable development, they engaged in fishing the major fisheries beyond their sustainable limits.

As is the case with many government programmes, the benefits of the government measures for fisheries development, were taken either by the rich fishermen or by the non-fishermen entrepreneurs. These people having easy access to the concerned government offices, ability to influence and pay bribe were able to take the benefits of subsidy, government loans, supply of accessories and other needs at concessional rates. The traditional fishermen were denied and deprived of these facilities to a large extent.

Most important fisheries are fully exploited or have reached the level where there exists a threat of resource depletion. Hence, competition between small-scale and large-scale fishermen occurs when both types of fishermen exploit the same fishing grounds. As the state of resource depletion occurs, fishermen adopt fine mesh nets, bull pulling trawlers and other destructive fishing methods, which meet their short term needs at the expense of long-term interest in resource sustainability. This response to resource depletion is most likely to harm the small-scale fishermen who have no alternative employment opportunities.

Many mechanized vessels fish close to the coasts and destroy the seeding grounds of shrimps, mackerels and other small crustacean families that cliché closely to the coastal waters for food, breeding. The traditional fishermen complain that the trawlers have exploited the shallow water fishing grounds where the traditional Ramponkars have been fishing for generations. The growing number of mechanized boats poses a threat to the very survival of the traditional fishermen. While the number of mechanized boats is increasing, the number of traditional crafts is decreasing. This shows that the mechanized crafts have come in the way of the livelihood of the traditional fishermen. Mechanization resulted in the increased catch, earnings and profit in the initial years. But of late, the catch is increasing very slowly. However, the fishermen are surviving because of rising prices of fish in the market. Mechanization has made very little positive impact. But its negative impact is more pronouncing. Fish resource is getting depleted over the years. Goa has a coastline of 105 km length. But there are 1134 trawlers, which are more than 4 times limit suggested by FAO (i.e., 30 trawlers per 10 kms of coastline).

The fishermen raise finance from two sources, namely, internal finance and external finance. The fishermen borrow from the commercial banks, co-operative banks, friends and relatives. But due to falling fish catch, they face the problem of repayment of loans.

There is a lack of efficient and systematic marketing system. As a result, the fishermen's share in consumer's rupee is around 60 per cent and remaining 40 per cent goes to middlemen. The price of fish is determined by the interaction of demand and supply conditions both at the landing centres and at the retail markets. The fish marketing in Goa is not fully developed on modern lines. At landing centres, fishes are disposed by auctioning. This provides maximum competition among buyers and enables quick disposal. Usually, at landing centres fish is not sold in weight because of the practical difficulties involved in the handling of such a perishable commodity. Hence, the sales are carried out by the measures of heaps or lots of different sizes.

The fishermen play an important role in Goa's fishery sector, particularly in the marketing part of it. With the growing mechanization, some men have moved into marketing niches at the jetties as middlemen, auctioneers or bankers. Often the husbands and the brothers-in-law of the fisherwomen purchase fish by auction at the jetties and give to these female vendors to sell in the markets of their destination. Sometimes, they themselves bargain directly at the jetty and either purchase fish for their own business or resell to other women. They face a number of problems in the market place such as non-availability of sufficient and proper space, payment of heavy tax, tough competition, and customer problem like too much bargaining, fights with other fisherwomen, lack of basic facilities like water, dustbin, toilet etc. at the market place.

The conflict between the traditional and mechanized fishermen is a conflict of interests, livelihood and environment. The conflict has given rise to open confrontation and litigation. It has also created public awareness, which has also pressed the need for a national fishing policy. It has opened a debate on the need of a uniform monsoon ban all over the country. It also supports the need for a sound fishing management policy so that there is plenty of fish for all the major contenders in the sea.

Sustainable fishing demands a sound fisheries management policy. It implies the rational exploitation of fisheries. It involves allocation of fish resources, formulation of rules and their enforcement. In short, there must be an effective fishery management based on the notion of sustainability, efficiency and equity through effective monitoring, control and surveillance measures. In the absence of sound fisheries management, there will be a tragedy of the commons. The fishermen have to incur heavy costs on the crafts, nets ad other accessories. The cost of Labour is also increasing. Therefore, they borrow from the financial agencies.

The common problems in marine fisheries production are:

(1) Over fishing and extensive fishing in shallow and near shore areas resulting in depletion of catch.

(2) Conflicts between traditional and mechanized vessels resulting in frequent clashes and loss of fishing gears.

(3) Entry of chartered and joint venture vessels and their hi-tech fishing which depletes the stock and diminish the catch of the local mechanized vessels, resulting in economic loss to the small operators.

(4) Monsoon fishing of brood stock, which come near shore for breeding, which reduces the fish stocking in commercially important species.

(5) Juvenile fishing particularly in their breeding grounds and ingress points in estuary and mangroves, which affects the marine fisheries seriously.

(6) Post harvest losses due to lack of proper landing, sorting, icing, storage, drying and lack of market facilities.

(7) Environmental and habitat degradation and damage resulting in loss of fisheries.

Compounding these are the escalating fuel cost, socio-economic problems etc. affecting the fishermen engaged in the industry.

Enforcement or implementation of some of the following measures would improve the situation and enhance the sustained fish production on a long time basis.

1. As the near shore catches are nearing sustainable yield limits, there is no point in increasing the fishing vessels in Goa. Exploitation of deep sea resources beyond 200m depth for which repair of old vessels, modernization of craft and gear and provision of scientific fish stock data to fishermen can be implemented.

2. Prohibition of mechanized trawlers in the 5 km area throughout the year.

3. Complete prohibition of fishing during monsoon.

4. Prohibition of mechanized fishing in inshore waters and estuaries.

5. Mesh size (24 mm) regulation to prevent juvenile fishing.

6. Sea ranching to be taken up to restore population density and artificial stocking.

7. All fishermen and entrepreneurs involved in fisheries should do responsible fishing as per laid down regulations.

SUGGESTIONS

(1) Throughout the west coast, the major problem for fishermen during the monsoon is loss of fishing days due to bad weather. There is no alternate employment opportunity during this period. Hence, it is essential to find out alternate employment potential. For the coastal area, the most suitable scheme is the development of brackishwater aquaculture. Other integrated projects can also be formulated with proper institutional arrangement for the sustainable supply of required inputs.

(2) Since the quantity of fish produced during monsoon is very low, it reduces the average household income of the fisher folk and consumers do not get sufficient quantity of fish at reasonable price. It provides opportunity to middlemen to exploit the situation for their own benefit. The major step to be taken to overcome this is the proper distribution of the fish produced during peak season. This can be done only by establishing adequate storage and processing facilities at all major landing centres. Usually, there will be glut in the market during peak season atleast for some species every year. In such cases, the fishermen dispose off the fish at throw-away price. This can be avoided by establishing a public agency to purchase the fish at minimum support price and the fish can be disposed off during monsoon after the required

processing. Such an agency must arrange adequate distribution system so that the fish can be transported and marketed at interior places. This will help the producer to get a remunerative price and the consumer the fish at a reasonable price.

(3) During monsoon the fishermen face difficulties not only in actual fishing operation but also in landing the catch. Most of the rural landing centres do not have proper jetty facilities. At many centres, fishing can be done even in monsoon if properly designed jetties are constructed.

(4) Goa's traditional fishing industry can best be developed through an integrated rural development program. Get local people involved.

(5) Attention should be given to shore facilities and adequate marketing and distribution infrastructure. Education, training and other forms of social investment should be an integral part of any small scale fisheries development.

(6) Choice of an intermediate technology like outboard motors, which are cheaper than trawlers and purse seines and highly fuel efficient will go a long way in the preservation of our coastal marine ecology as well as prevent the displacement of the existing traditional labour force.

(7) Active participation of the fishing community through their own traditional organizations, in planning, and implementation of development activities is needed, if they are to be successful.

(8) The important role of our fishermen in fish processing and fish curing, primary health care and nutrition should also be taken into account. A supplementary alternative source of income and employment for fishermen, like fish farming should be designed to counter occasions of scarcity of marine fish resources.

(9) Improvement and modification of locally developed fishing gear like canoes and nets should be made, if we

are to respect the age-old wisdom of the Goa's Ramponkar Communities.

(10) Goa has good facilities for marshy land and brackish water pisciculture. This can provide employment to our Ramponkar community. Traditional fishing communities continuously face potential hazards and threats of losing their lives during fishing. In this respect a consolidated insurance scheme for their next of kin could ensure security for their dependents.

(11) Fish finders should be banned because the fisherman can detect fish with it from a long distance and he takes away the full stock in that particular place, which is detrimental for sustainable development.

(12) Those who break / violate the ban should be punished and their licenses should be cancelled.

(13) The fishermen must be provided information about prices of various species in the international markets. Otherwise the agents exploit the fishermen and buy the fish from them at a low price.

(14) Traditional fishermen should be encouraged to continue fishing by provision of efficient crafts and gears and by improving marketing system.

(15) Low cost motorization must be developed for the benefit of small scale fishermen.

(16) Mechanical fishing should be diversified. Most of them concentrate on shrimp production causing fear of overexploitation. Hence, diversified deep sea exploitation should be encouraged.

(17) Handling, processing and distribution is also important in marine sector and therefore, necessary infrastructure facilities must be developed to efficiently carryout the above things.

(18) Thrust on development and marketing of edible fish products from low species must also be given. The handling, processing and marketing of low species

hither to ignored must be improved. Adequate capital assistance, subsidies, tax/ duty exemptions etc. should be extended for the same in the future for at least five years. Foreign companies can also be encouraged for joint venture.

(19) Infrastructural facilities such as fishing harbours, landing centres, ice plants, cold storages, processing plants, maintenance and repair yards etc are the main types of infrastructural facilities required by the marine fishing industry. Fishing industrial estates, ancillary and support services are to be developed to improve the fishery sector.

(20) Financial support from term lending institutions must be extended in time to invest in deep sea fishery as well as inshore fishing.

(21) Maintenance of up-to-date data would help in careful analysis of the fishery sector which would enable the enforcement of measures to prevent poor fishing and to ensure optimal resources exploitation.

AREAS FOR FURTHER RESEARCH

(1) The study of fisher folk community is one area, which requires immediate attention. In Goa, such studies are lacking.

(2) Mechanization and displacement is another area, which need to be researched. The people engaged in allied occupations of fishing are displaced.

(3) Proper sociological analysis of the process of managing fisheries is necessary. The study of social impact is a must as they can jeopardize the success of fisheries interventions. Gender, age, community, education, household etc. are important social issues, which require sociological analysis in the context of fisheries system.

(4) Economics of shrimp farming in Goa.

(5) An in depth study of productivity, profitability and efficiency measures of mechanized crafts in Goa.

(6) Price fluctuations of various species of fish and factors responsible for the same.

(7) Marine fish marketing and impact of price changes on marine fish landings in Goa.

(8) Role of fishermen's co-operative societies in sustainable fishery development.

(9) Price trends and marketing efficiency of inland fish in Goa.

(10) Social and ecological problems of shrimp farming.

Bibliography

BOOKS

Abidi, S. A. et. el. (1999). Vision of Indian Fisheries of 21st Century. Mumbai: Central Institute of Fisheries Education.

Agarwal, S.C. (1989). Fishery Management. New Delhi: Ashish Publishing House.

Alagaraja, K. et. al. (1992). Marine Fish Production of Maritime States of the West Coast of India: Prospects, Problem and Management. Cochin: Central Marine Fisheries Research Institute.

Allsopp, W.H.L. (1985). Fishery Development Experiences. England: Fishing News Book Limited.

Alvares, Claude (2002). Fish Curry and Rice. Goa. Mapusa: The Goa Foundation Publication.

Ananth, P.N. (2000). Marine Fisheries Extension. New Delhi: Discovery publishing house.

Anderson, Lee G. (1977). The Economics of Fishery Management. London: John Hopkins University Press.

Andrew, P. (1999). Economics of Brackish water Shrimp culture. New Delhi: Daya Publishing House.

Asada Y.,Hirasawa Y.and Nagasaki F.(1983) Fishery Management in Japan. FAO. Fisheries Technical Paper, No.238.

Angle, Prabhakar. (2001). Goa- An Economic Update. Bombay: Goa Hindu Association Kala Vibhag.

Bailey, C. (1982). Small-scale Fisheries of San Miguel Bay, Philippines: Occupational and Geographic Mobility, ICLARM Technical Report No. 10. Manila: International Centre for Living Aquatic Resources Management; Institute of Fisheries Development and Research, College of Fisheries, University of the Philippines in the Visays; and the United Nations University, 56 pp.

Bailey, C.(1983).The Sociology of Production in Rural Malay Society.New York and Kuala Lumpur: Oxford University Press, pp 226.

Bailey, C. (1987). Government protection of traditional resource use rights: the case of Indonesian fisheries. Pp. 292-308. In: C.C. KORTEN (ed.), Community Management: Asian Experience and Perspectives. Hartford, Conn.: Kumarian Press.

Bal, D.V. and Rao Virabhadra K. (1984). Marine Fisheries. New Delhi: Tata Mc graw Hill Publishing Company Limited.

Barduch, John. (1969). Harvest of the Sea. London: George Allen Unwin Ltd.

Barnable, Gilbert and Barnable Quet Regine. (1997). Ecology and Management of Coastal Waters: The Aquatic Environment. Chiechester: Praxis Publication.

Bavinck, Maarten. (2001). Marine Resource Management – Conflict and Regulation in the Fisheries of the Coromandel Coast. New Delhi: Sage Publication.

Bhattacharya, Hrishikes. (2002). Commercial Exploitation of Fisheries. New Delhi: Oxford University Press.

Bhatta, Ramachandra .(1996). 'Role of Co-operation in the Management and Development of Marine Fisheries of Coastal Karnataka', in Rajgopalan (eds.): Rediscovering Co-operation: strategies for the models of tomorrow. Anand: Institute of Rural Management.

Brown,C. Robert. (1998). 'Community –Based Cooperative Management: Renewed interest in an old paradigm', in Tony J Pitcher et al (eds.): Reinventing Fisheries Management. London: Kluwer Academic Publishers.

Buckwork, Rilk C. (1998). 'World Fisheries are in Crisis? We must respond!,' in Tony J Pitcher et. al. (eds.): Reinventing Fisheries Management. London: Kluwer Academic Publishers.

Charles, T. Anthony. (1998). 'Beyond the Status quo: Rethinking Fishery Management', in Tony J Pitcher et. al. (eds.): Reinventing Fisheries Management London: Kluwer Academic Publishers.

Chaudhari, M.R. (1975). Economic Geography. New Delhi: Oxford and IBH Publishing Co.

Chaudhary P. D. (2002) "Role of Women in Fisheries Development in Gujarat State" in "Women in Fisheries – Indian Society of Fisheries Professionals"

Cherunilam, Francis. (1993). Fisheries Global Perspective and Indian Development. New Delhi: Himalaya Publishing House.

Dahmani, M. (1987). The Fisheries Regime of the Exclusive Economic Zone. Dordect: Martinus Nizhofe Publishers

Dehadrai, P.V. (1982). 'Traditional and Modern Fisheries: An Integrated Approach', in Tridevi (eds.):Proceedings of International Conference on Deep Sea Fishing. New Delhi.

De Souza Teotonio R (ed) (1990), "Goa through the ages- Vol II An Economic History, Concept Publishing Company, New Delhi.

Devaraj, M. (1996). 'Coastal Biodiversity: Conservation and Sustainable Management', in Menon N.G. and Pillai.C.S.G. (eds.): Marine Biodiversity: Conservation and Management. Cochin: Central Marine Fisheries Research Institute.

ECOFORUM. (2000). Fish Curry Rice: A Citizens Report on the State of the Goan Environment. Mapusa: The Other India Press.

Giriappa, S. (1994). Role of Fisheries in Rural Development. New Delhi: Daya Publishing House.

Giriappa, S. (1994). Rural Development in Action. New Delhi: Daya Publishing House.

Goa Chamber of Commerce and Industry, (1990). Industrial and Commercial Directory, Goa.

Gracy, M.M. (1988). 'Impact of Technological Advancement on Socio-Economics of Women in Fisheries, Kerala', in M.S. Hameed and B.M. Kurup(eds.): Technological Advancement in Fisheries. Cochin: Cochin University of Science and Technology.

Haggan, Nigel. (1998). 'Reinventing the Tree: Reflections on the Organic Growth and Creative Running of Fisheries Management Structure', in Tony J. Pitcher et. al. (eds.): Reinventing Fisheries Management. London: Kluwer Academic Publishers.

Hameed M. S. and Kurup B. M. (eds.) (1998). 'Technological Advancements in Fisheries', Publication no. 1 – School Ind 1. Fish Cochin : Cochin University of Science and Technology.

Hannesson, Rognvaldur (1981). "The Contribution of Economics to Canadian Fisheries Management". In : Anderson, Leegh. (1981). Economic Analysis for Fishery

Hannesson, R. (1993). Bio-Economic Analysis of Fisheries. England: Fishing News Book. Hannessson, R. (1996). Fisheries Management: The Case Study of the North Atlantic Cod. London: Fishing News Book.

Hariri, Khaled. (1985). Fisheries Development in the North West Indian Ocean – The Impact of Commercial Fishing Arrangement. London: Ithaca Press.

Hawthorne Edward P. (1978). 'The Management of Technology', U.K: Mc Graw Hill Book Co. Ltd., Pg- 1

Ibrahim, P. (1992). Fisheries Development in India. New Delhi: Classical Publishing Company.

Jackson, Roy I. and Royce William. (1986). Ocean Forum: An Interpretative History of The International North Pacific Fisheries Commission. England: Fishing News Book Limited.

Jhingran, V.G. (1991). Fish and Fisheries of India. Place: Hindustan Publishing Corporation (India).

Joseph, Mohan M (ed.) (1988). The First Indian Fisheries Forum Proceedings, Asian Fisheries Society, Indian Branch, Mangalore, pp. 433-438.

Judicello, Suzanne et. al. (1999). Fish Market and Fisherman (The Economics of Four Fishing). U.S.: Island Press.

Kay, Robert and Alder, Jacqueline (1999). Coastal planning and Management. New York: E & FN SPOW.

Kochery, Thomas. (1998). Indian Fisheries over 50 years. Bangalore: Books for Change.

Koli, Singh M.P. and Tewari Ratna. (eds). (2002). Women in Fisheries. Mumbai: Indian Society of Fisheries.

Korakandy, R. (1996). Economics of Fisheries Management. New Delhi: Daya Publishing House.

Korakandy, R. (1999). Technological Change and Development of Marine Fishing Industry in India: A case study of Kerala. New Delhi: Daya Publishing House.

Krishnamurthi B. et.al (1995), 'Fishery Management', Proceedings of the seminar on Fisheries: A Multibillion Dollar Industry, Held at Madras, from August 17 to 19, 1995.

Kurien, John and Sebatian Mathew (1982) "Technological Change in Fishing - Its Impact on Fishermen", (Mimeo) Centre for Development Studies, Trivandrum. Pp. 12-13.

Kurien, John. (1991). Running the Commons and Response of Commoners: Coastal Overfishing and Fishermen Actions in Kerala State. India. United National Research Institute for Social Development.

Kurien, John, (2002). People and the Sea: Tropical Majority World Perspective. Netherlands: Netherlands Institute for the Social Science and Center for maritime Research (MARE).

Kurien, John. (1998). Property Rights, Resources Management and Governance: Crafting an Institutional Frame Work for Global Marine Fisheries. Thiruvanthpuram: Center for Developmental Studies and South Indian Federation of Fisheries Societies.

Laevastu, Taivo and Hayes L. Marray. (1981).Fisheries Oceanography and Ecology. England: Fishing News Book Limited.

Mchoowin. James R. (1990). Crisis in the world Fisheries: People, Problem and Policies. California: Stanford University Press.

Panikkar, K.K.P. et. al. (1998). 'Structural Change in the Traditional Fishery of Kerala and Its Socio Economic Implications', in M .S. Hameed and B. M. Kurup (eds.): Technological Advancements in Fisheries. Cochin: Cochin University of Science & Technology.

Pauly, D. (1979) Theory and management of tropical multispecies stocks: a review, with emphasis on the Southeast Asian fisheries. ICLARM Studies and Reviews 1. Manila: International Center for Living Aquatic Resources Management, 35 pp.

Pauly, Daniel et. al. (1998). 'Speaking for themselves, new acts, new actors and a new deal', in Tony J. Pitcher et. al. (eds.): Reinventing fisheries management. London: Kluwer Academic Publishers.

Pillai, V. N. and Menon N.G. (2000). 'Marine Fisheries Research and Management', Cochin: CMFRI.

Pitcher, J. Tony and Pauly Daniel. (1998). 'Rebuilding Ecosystem, Not Sustainability, as the Proper Goal of Fishery Management', in Tony J Pitcher et. al. (eds.): Reinventing Fisheries Management. London: Kluwer Academic Publishers.

Porter, Gareth. (1998). 'Fisheries Subsidies, Overfishing and Trade', Place: United Nations Environment programme.

Pramanik, S.K. (1993). Fishermen Community of Coastal Villages in W. Bengal. New Delhi: Rawat Publication.

Rajgopalan, R. (eds.). (1996). 'Voices for the Ocean: A Report to the Independent World Commission on the Ocean', Chennai: International Ocean Institute.

Rana K. and V. Q. Perrez-Corral(1998), "Where are women in fisheries?", Fifth Asian Fisheries Forum, Asian Fisheries Society, International Symposium on Women in Asian Fisheries, published by ICLARM – The World Fish Centre, Malaysia.

Rao, Hanumantha C. H. & Joselin P. C. (ed). (1979). 'Reflection on Economic Development and Social Change', New Delhi: Allied Publishers Pvt. Ltd. Pg- 63.

Rao, K. M (1998). Status of Fisheries Extension and Information Service in India. In Mohan Joseph M. (Ed.) 1998. Research and Development in Marine Fisheries. Proc. of one First Fisheries Forum. Asian Fisheries Society, Indian Branch, Mangalore, India.

Rao P.S. (1983) Fishery Economics and Management in India, Pioneer Publishers and Distributors, Mumbai Ass. India 34 (1& 22): 18-25.

Rao, Subba N. (1986). Economics of Fisheries. New Delhi: Daya Publishing House.

Roy, Prithwish. (1997). Economic Geography: A Study of Resources. Calcutta: New Central Book Agency (P) Ltd.

Ruddle,K.(1984).'The continuity of Traditional Management practices: The case of Japanese coastal fisheries' In: K Ruddle and R.E. Johannes).(Eds) The Traditional Management of Coastal Systerms in Asia and the Pacific, Unesco,Paris.

Sahrhage, Dictrich and Lundbeck Johannes. (1992). Ministry of Fisheries. New York: Springer Verlag.

Sakthivel, M. (2001). 'Challenging Task for Sustainable Fisheries Development in the Millennium 2000', in T.J. Pandian (eds.): Sustainable Indian Fisheries. New Delhi: National Academy of Agriculture Science.

Sall, Aliou et. al. (2000). A Trilogue on Power, Intervention and Organisation in Fisheries. Tamil Nadu: International Collective in Support of Fishworker.

Sathiadhas, R. and A. Kanagam (2000). Pillai V.N. and N.G. Menon (eds). Marine Fisheries Research and Management Cochin: CMFRI.

Sathiadhas, R. (1997). Production and Marketing Management of Marine Fishing in India. New Delhi: Daya Publishing House.

Scott, Anthony. (1998). 'Cooperation and Quotas', in Tony J Pitcher et. al. (eds.): Reinventing Fisheries Management. London: Kluwer Academic Publishers

Singh, Jagbir. (2005). 'Environmental Development Challenges and Opportunities', New-Delhi: I. K. International Pvt. Ltd.

Somvanshi, V.S. (2001). 'Problem and Prospects of Deep Sea Fishing', in Pandian T.T.(eds.): Sustainable Indian Fisheries. New Delhi: National Academy of Agriculture science.

Srinath, M.(2003). 'An Appraisal of the Exploited Marine Fishery Resources of India'. In: Joseph M.Mohan and Jayaprakash (eds.) "Status of Exploited Marine Fishery Resources of India, Kochi, Central Marine Fisheries Research Institute.

Srivastava, U.K. et.al. (1986). Impact of Mechanization on Small Fishermen: Analysis and Village Studies. New Delhi: Concept Publishing Company.

Subramanium, S. (2001). 'Status and Prospects of Fisheries Production and Processing in Goa', Paper presented in the1st Symposium on Post-Harvest Technologies for Agriculture Production and Prospects for the Food Processing Industry in Konkan Region.23-24 November, International Center, Goa.

Sutton, Michal. (1998). 'Harnessing Market Forces and Consumer Power in Favour of Sustainable Fisheries', in Tony J Pitcher et. al. (eds.): Reinventing Fisheries Management. London: Kluwer Academic Publishers.

S. Suwanrangsi (1998), "Women's role in fisheries", Fifth Asian Fisheries Forum, Asian Fisheries Society, International Symposium on Women in Asian Fisheries, published by ICLARM – The World Fish Centre, Malaysia.

Symes, David (eds). (1998). Property Rights and Regulation System in India. London: Fishing News Book.

Venugopal, S. (2005). 'Aquaculture: Jaipur : Pointer Publishers.

Villafuerte,E.D.and C.Bailey. (1982) Systems of sharing and patterns of ownership. In: C.Bailey (ed),Small Scale Fisheries of San Miguel Bay, Philippines: Social Aspects of Reports of production and marketing ICLARM Technical No.9.Manila:International Centre for Living Aquatic Resources Management; Institute of Fisheries Development and Research, College of Fisheries ,University of the Philippines in the Visayas:and the United Nations University.

Yamamoto T. and Shot K. (1992) (eds.) International Perspectives on Fisheries Management: Proceedings of the JIFRS IIFT/ ZENGYOREN Symposium on Fisheries Management, Tokyo, 26 August-3 September 1991.National Federation of Fisheries Cooperative Associations (ZENGYOREN) in association with Japan International Fisheries Research Society (JIFRS).

ARTICLES

Abidi, S.A.H. (1981). 'Status of small-scale fisheries in the Union Territory of Andaman and Nicobar Island', CMFRI bulletin, 30-B, pp.57-59.

Abraham, Anita. (1985). 'Subsistence Credit: Survival Strategies Among Traditional Fishermen', Economic and Political Weekly, Vol.XX(6):247- 252.

Alam, S.K. and Rao Sabba P.V. (1998). 'Socio-Economic Condition of Traditional Fishermen of Vitava Village in Thane District', Journal of the Indian Fisheries Association, Vol.18, Pp.573-577.

Ananth, P.N. (2002). 'Fisheries Extension in India: A Boon for Emerging Technology', University News, Vol.38 (8):1-3.

Anon, (1998). Faster Fisheries Development: Consultancy Infrastructure is the Key. Fishing Chimes 17 (2), pp. 5-7.

Azhar, T. (1980). Some preliminary notes on the by-catch of prawn trawlers off the west coast of Peninsular Malaysia. Pp. 64-69. Report of the Workshop on the Biology and Resources of Penaeid Shrimps in the South China Sea Area- Part 1. SCS/GEN? 80/26. Manila: South / China Sea Fisheries Development and Coordinating Programme.

Bailey, C. (1982). Small-scale Fisheries of San Miguel Bay, Philippines: Occupational and Geographic Mobility, ICLARM Technical Report No. 10. Manila: International Centre for Living Aquatic Resources Management; Institute of Fisheries Development and Research, College of Fisheries, University of the Philippines in the Visays; and the United Nations University, 56 pp.

Bailey, C. (1985). The blue revolution: the impact of technological innovation on Third World Fisheries. The Rural Sociologist 5(4): 259-66.

Bailey, C. and Zerner C. (1991). "Role of traditional fisheries resource management system for sustainable resource utilization", paper presented at the Fisheries Forum of Centre for Research Institute of Fisheries, Sukabumi, West Java.

Barbosa, M. Alexander. (2000). 'Fishing for High Living', Goa Today, Vol. XXXVII(4):10-14.

Bavinck, Maarten. (1997). 'Changing Balance of Power at Sea', Economic and Political Weekly, Vol. XXXII(5): 10-14.

Bhatta, Ramachandra and Bhat, Mahadev (1998). 'Impact of Aquaculture on the Management of Estuaries in India', Environmental Conservation -An International Journal of environmental science, Vol25(2):109-121.

Bhatta, Ramachandra et. al. (2000). 'An Economic analysis of Fishing Operation in Coastal Karnataka', Journal of Social and Economic Development, Vol.II(2):329-347.

Bhatta, Ramachandra. (2000). 'Estimating Socio-Economic Impact of Alternative Fishery Management Regulation', Coastine: A Coastal Policy Research News Letter, No. 2, pp.9-11.

Bhatta, Ramachandra. (2003). 'Proceedings of the Workshop on Fishing Ban, Fish Famine and Livelihood Issues', Organised by Shri Dharmasthala Manjunatheshwar College of Law and Center for Environmental Education, Research and Advocacy, Mangalore. At S.D.M.National Law School of India.

Bhatta Ramachandra and K. Aruna Rao. (2003), "Women's Livelihood in Fisheries in Coastal Karnataka, India", Indian Journal of Gender Studies, Vol. 10. No. 2.

Bierhuizen, B.R. (2000). 'Fishing Pattern and Intensity in Kanyakumari District', Bay of Bengal Programme News.

Chidambaram, K. (1973). 'Infrastructure for Fishing Industry', Proceeding of the Symposium on Living Resources, Central Marine Fisheries Research Institute.

Chidambaram, K.(1991). 'Management and Conservation of Marine Fishery Resources', CMFRI Bulletin , 44, Part Three, Pp.594-603.

Davis, Sandy. (2001). 'Fisheries Management- Fish Code Report of the National Workshop on Fisheries Monitoring Control and Surveillance' in support of Fisheries Management held in Goa during 12-17 February, 2001. pg. 11-20.

D'souza, Joe. (2002). 'Fishing Woes to the Fore', Goa Today, Vol. VII(4): 14-18

Edeitrand, Drewes. (1982). 'Three Fishing Villages in Tamil Nadu', A Socio -Economic Survey with Special Reference to Role and Status of Women, Bay of Bengal Progremme News. /10p/14.

FAO. (1961). 'Cost and Earning Investigation of Primary Fishing Enterprise', FAO Fisheries Technical Paper, No: 10.

FAO. (1982). 'Management Concepts of Small Scale Fisheries: Economic and Social Aspects', FAO Fisheries Technical Paper, No. 228.

FAO. (1989). 'Administration and Conflict Management in Japanese Coastal Fisheries', FAO Fisheries Technical Paper, No. 273.

FAO. (1993). 'Marine Fisheries and law of the Sea; A Decade of Change', Special Chapter (revised) of The State of The Food and Agriculture 1992, FAO Fisheries Department.

FAO. (1999). 'Economic Viability of Marine Capture Fisheries Finding of a Global Study and an Interregional Workshop', FAO Fisheries Technical Paper, No. 37.

FAO. (2000). 'Report on the National Workshop on Fisheries Monitoring Control and Surveillance in Support of Fisheries Management', Goa, 12-17 Feb.

FAO. (2001). 'Managing Fishing Capacity: A Review of Policy and Technical Issues', FAO Fisheries Technical Paper, No. 409.

FAO. (2001). 'Techno- Economic Performance of Marine Capture Fisheries', FAO Fisheries Technical Paper, No. 421. .

Fernando, Ambrose. (1981). 'Community Development and Infrastructure Facilities for Improving the Socio-Economic Condition of Fishermen', CMFRI bulletin, 30-A, Pp.53-56.

Freire, Juan and Garcia Antonio. (2000). 'Socio-Economic and Biological Causes of Management Failure in Europe Artisanal Fisheries: The Case of Galicia (New Spain)', Marine International Journal of Ocean Affairs, Vol. 24(2):375-389.

Gopalkumar, K. (1998).'Fish Products for Gainful Employment for Women', Proceedings of All India Workshop on Gainful Employment for Women in Fisheries Field, March 7-8, Cochin, Department of Science & Technology.

Goss, Jasper (1998). 'Conflict and Resolution in Indian Shrimpaquaculture', Economic and Political Weekly, Vol. XXXII(8):383-384

Govindan, T.K. (1973). 'Application of Technology in Optimum Utilization of the Fishery Resources in India', Proceedings of the Symposium on Living Resources, Central Marine Fisheries Research Institute, Cochin.

Gupta, Pratap Ram (1995). 'Management of Fisheries Development', Economic and Political Weekly, Vol. XXII(37):1910.

Hameed, Shahul M. Mukundan M. (1991).'Strategies for the development and management of purse seine fishing in India', CMFRI bulletin, 44, Part Three, Pp. 643-646.

Hapke, M. Holly (2001). 'Development, Gender and household Survival in Kerala Fishery', Economic and Political Weekly, Vol. XXXVI.(13):1103-11

Immunel, Sheela. (1975). 'Training Programme for Fisherwomen on Preparation of Food from Seaweed-An Evaluation Study', Marine Fisheries Information Series, No. 137, Pp. 3-7.

Jacod,T. et. al. (1979). 'Socio-Economic Implication of Purse Seine Operation in Karnataka', Marine Fisheries Information Service, No. 12.

James, P.S.B.R. (1992). 'The Indian Marine Fisheries Resources Scenario; Past, Present and Future', Indian Journal of Fisheries.Vol. 39(12):1-8.

James, P.S.B.R. and Alagaraja K. (1998). 'Assessment of Marine Fisher Resources of India' . Proceedings of National Workshop on Fishery Data and Fishing Industry. 14-15 Oct. Vishakapathnam. Fishery Survey of India.

James, P.S.B.R.(1991). 'Strategies for marine fisheries development in india', CMFRI bulletin, 44, Part Three, Pp. 511-524.

Jhingran, Arun G. and Halder DD. (1998). 'Researcher, Appropriate Technology and Role of Women in the Development of Fisheries', Proceedings of Workshop on All India Employment for Women in

Fisheries Field, 7-8 March, Cochin, Department for Science & Technology.

Johnson, Deepak. (2001). 'Wealth and Waste: Contrasting Legacies of Fisheries Development Since 1950s', Economic and Political Weekly, Vol. XXXVI(13):1095-1101.

Kagoo, Emerson and Rajalakshimi N. (2002). 'Environment and Social Conflict of Aquaculture in Tamil Nadu and Andhra Pradesh', Journal of Social and Economic Development, Vol. IV (1):13-26.

Kemparaju, S. (1994). 'Drift Gillnet Fishery of Goa', Marine Fisheries Information Series, No.128. pp.5-8.

Koli N.D. (2001). 'Fish workers Demand', Economic and Political Weekly, Vol. XXXVI (1):80.

Korakandy, Ramakrishna. (1996). 'Managing World Fisheries: Third World Loss', Economic and Political Weekly, and Vol. XXXI (34):2289-2291.

Korakandy, Ramakrishna. (2001). 'Aquarium Reforms: Will Actual Fishermen Benefit', Economic and Political Weekly, Vol. XXXVI (38):3590-3591.

Kumar, R.S.P. (2001). 'New technology and Artisanal Fishermen in Kerala', Indian Journal of Labour Economics, Vol.44 (1):75-84.

Kurien, John and Paul Anthony. (2001). 'Social Security Nets for Marine Fisheries', Working Paper, 318, Central Institute for Development.

Kurien John, (1978). Entry of Big Business into Fishing: Its impact on the fish economy, Economic and Political Weekly, 13 (36), pp. 1557-1965.

Kurien, John. (1982). 'Technological Change in Fishing: Its Impact on Fishermen', Center for Development Studies, Cochin.

Kurien, John. (1995). 'Impact of Joint Venture on Fish Economy', Economic and Political Weekly, Vol. XXX (6):300-302.

Kurien, John. (2000). 'Factoring Social Cultural Dimensions in Food and Livelihood Security Issues of Marine Fisheries: A case Study of Kerala States', Working Paper. 299, Center for Development Studies, Cochin.

Kurien, John. (2000). 'Icelandic Fisheries Governance: A Third World Understanding', Economic and Political Weekly, Vol. XXXV (34):3061-3066.

Kurien, John. (2000). 'Responsible Fisheries: Can it be Achieved with a Code of Conduct', Bay of Bengal Programme News, No.11, Pp. 10-14.

Lamin, Bin Selehan. (20003). 'Situation of MCS in Malaysia', Proceedings of Workshop on Monitoring Control and Surveillance (MCS) for Marine Fisheries Management in Goa, 12-17Feb, FAO / Norway Co-operatives Programme.

Leon, Charles (2003). 'Human Rights Violation Against Fisherwomen in Kerala', Loyala Journal of Social Science, Vol. XVIII (1):71-95.

Luther, G. et. al. (2001). 'Gillnet Fisheries of India', Marine Fisheries Information Series, No. 150, Pp.1-24.

Mammen, T.A. (1991). 'Management and Conservation of Marine Fishery Resources', CMFRI bulletin, 44, Part Three, Pp.604-612.

Mansveta, Deb. (2001). 'The Sustainable Contribution of Fisheries in Southeast Asia', Keynote Addresses, 5th and 6th Asian Fisheries Forum (Philippines), Asian Fisheries Society.

Martosu, Broto Purwito.(2000). 'Some Notes on Fisheries and Coastal Fisheries Management in South and Southeast Asia', Proceedings of Workshop on Sustainable Fisheries Management in Goa, NIO, 11-12 October, Dona Paula.

Mathai, John. (1998). 'Entrepreneurial Development and Self Employment Programme Among Women and Role of Industries Service Institute', Proceedings of All India Workshop on Gainful Employment for Women in Fisheries Field, March 7-8, Cochin, Department of Science & Technology.

Menon, A.G.(1991). 'Marine National Park and Conservation of Fisheries Resources', CMFRI bulletin,44, Part Three, Pp. 668-672.

Menon, Meena. (2002). 'Tawa Matsya Sangh: Fishing for their lives', The Hindu Survey of the Environment, Pp. 99-103.

Mishra, Snehasish. (1998). 'Perspective of Aquaculture Development', Yojana, Vol. 42(3):37-41.

Modayil, Joseph Mohan. (2002). 'Marine Fisheries in a Crucial Phase', The Hindu Survey of Agriculture, Pp. 139-142.

Mohanta, K.N. (1997). 'Formation of Feeds for Giant Freshwater Prawn Larvae as Alternative or Additions to Artemia Salina: A Review', Fishing Chimes, Vol. 17(7):11-12.

Mohanta, K.N. and Subramaniam S. (1997). 'Problem and Perspective of Freshwater Prawn Farming in India', Fishing Chimes, Vol.17 (9):11-12.

Mukual. (1993). 'Traditional Fisher-People against Fishing Harbour', Economic and Political Weekly, Vol. XXVIII (38):1974-1975.

Mukual. (1994). 'Aquaculture Boom, Who Pays', Economic and Political Weekly, Vol. (49):3075-3078.

Nair Mini. (1998), "Women in Fisheries – Emancipation through Co-operatives" in M.S. Hammed and B. M. Kurup (Eds.) Technological Advancements in Fisheries, Publn. No. - 1 – School Indl. Fish, Cochin University of Science and Technology, Cochin.

Organisation for Economic Co-operation and Development. (1997). 'Towards Sustainable Fisheries: Economic Aspect of Management of Living Marine Resources', OECD.

Panayotou, T. (1980). Economic conditions and prospects of small-scale fishermen in Thailand. Marine Policy 4(2): 142-146.

Panayotou, Theodore. (1982). 'Management Concepts for Small-Scale Fisheries: Economic and Social Aspects', FAO Fisheries Technical Paper, No. 220, FAO, Pp.1-53.

Pandey, A.K. and Das P. (2000). 'Fish Biodiversity Conversion: Theory and Practice', Fishing Chimes, Vol. 22(5):14-24.

Pandey, S.K. (2002). 'Growth in Inland Fish Production of India During the Last Decade (1991-2002): An Analysis', Vol.IV(2):16-13.

Pangam, D.R. et. al. (1981). 'Industrial Development Bank of India and its Role in Financing the Fisheries Projects', CMFRI bulletin, 30-A, Pp.153-161.

Panikar K.K.P. et. al. (1994). 'An Economic Evaluation of Purse seine Fishery Along Goa Coast', Marine Fisheries Information Series, No127, pp.4-8.

Panikar, K.K.P. and Alagaraja K. (1981). 'Socio-Economic Status of Fishermen Community of Calicut Area', Marine Fisheries Information Series, No 53, pp.2-12.

Panikar, K.K.P. and Venketesh B. (1991). 'Role of NABARD in Financing Marine Fisheries Projects', CMFRI bulletin, 44, Part Three, Pp.579-583.

Phillip, Townsley. (1998). 'Social Issues in Fisheries', FAO Fisheries Technical Paper, No. 375, FAO.

Pulin, S.V. Roger. (eds). (1997). 'Towards Policies for Conservation and Sustainable use of Aquatic Genetic Resources', International Center for Living Aquatic Resources Management,

Rajasena, D. (2001). 'Technology and labour Process in Marine Fishery-The Kerala Experience', The Indian Journal of Labour Economics,. Vol.44 (12): 269-276.

Rajlakshimi,T. (1994). 'A Suggested Plan for Development of Marine Fisheries Sector of Andhara Pradesh', CMFRI bulletin, 44, Part Three, Pp.548-553.

Ramamurthy, S. (1981). 'Traditional Practices of Coastal Aquaculture and Sustenance Fishery in India', CMFRI bulletin, 30-A, pp31-36.

Rao, Janardhan D. (1988). 'Resource information Requirement of Fisheries Development Agencies', Proceedings of National Workshop on Fishery Resources Data and Fishing Industry, October 14-15, Vishakapatnam, Fishery of Survey of India.

Rao, P.S. and Pandey S.K. (1990). 'Cost of Production of Various Types of Mechanised Fishing at Versova Landing Center, Bombay', Proceedings of Second Indian Fisheries Forum, Mangalore, 27-31 May.

Rao, S.N. (1981). 'Public Policies and Planning of Rural Fisheries in Kerala', CMFRI bulletin, 30-A, PP.171-175.

Rao, Subba N. (1991). 'Role of Social Science Research in the Integrated Development and Management of Marine Fisheries', CMFRI bulletin, 44, Part Three, Pp. 612-619.

Rao, Subba N.(1982). 'Kolleru Lake, a Boon to Fishermen', Yojana, Vol. XXVI(20):27-23.

Rubinoff, Janet. (1999). 'Fishing for Status Impact of Development on Goa's Fisherwomen', Women study International Forum, Vol. 22(6):631-644.

Rubinoff, Janet. (2001). 'Pink Gold: Transformation of Backishwater Aquaculture on Goa's Khazan Land', Economic and Political Weekly, Vol. XXXVI (13):1108-1113.

Sagard, Guraraj et. al. (2002). 'Fish Processing Industries in Karnataka: An Economic Analysis', Proceedings of the Fifth Indian Fisheries Forum, AFSLB Mangalore and AOB Bhubaneshwar.

Samal, Kishor. (2002). 'Shrimpculture in Chilika Lake', Economic and Political Weekly, Vol. XXXVII (18):1714-1718

Sangamaheswaran, A.P. et. al. (2000). 'Bioremediation in Aquaculture', Sea Food Export Journal, Vol. XXXI (7):31-35.

Sardjono,I., Trawlers banned in Indonesia .ICLARM Newsletter 3(4):3.

Sathiadhas, R. (1994). 'Traditional Fishermen in Low Income Trap-A Case Study in Thanjavur Coast of Tamil Nadu', Marine Fisheries Information Series, No. 135, pp.5-10.

Sathiadhas, R. (1996). 'Economic Evaluation of Marine Fisheries in India for Sustainable Production and Coastal Zone Development', NAGA. Malaysia, the ICLARM Quarterly, Pp.54-56.

Sathiadhas, R. and Panikkar K.K.P. (1992). 'Share of fisherman and middlemen in consumer Price: A study at Madras Region', Journal of Marine Biological Association of India, No.34.

Sathidhas, R. (1997). 'Socio- Economic Structural Changes in the Marine Fisheries Sector of India and Coastal Zone Management', Proceeding of the Seminar on Coastal Zone Management, Nagacoil, Institute of Coastal Area Studies.

Sathidhas, R. and Venkataraman G. (1991). 'Impact of Mechanised Fishing on the Socio-Economic Condition of the Fishermen of Sakthikulangara-Neendakara, Kerala', Marine Fisheries Information Series, No.110, P.p. 1-18.

Sehara, D.B.S et. al. (1991). 'Economic Feasibility of Trawling in Maharashtra', Marine Fisheries Information Series, No. 114. Pp.8-15.

Sehara, D.B.S. (1991). 'Socio-Economic of Trawler Fishery in Saurastra- A Case study', Marine Fisheries Information Series, No. 110, pp.1-7.

Sehara, D.B.S. et. al. (1988). 'An Evaluation of Fishermen Economy in Maharashtra and Gujarat: A Case Study', Central Marine Fisheries Research Institute Bulletin, No. 44.

Shahajan, K.M. (1996). 'Deep Sea Fishing', Economic and Political Weekly, Vol. XXXI (5):263-266.

Shahajan, S. (1999). 'Integrated Marine Fisheries Development Projects, Kerala Experience', Yojana, Vol. 43(12):37-35.

Shakila, Jaya R. et. al. (2002). 'Development in Fish Processing Technology and Export Potential for Fish and Fishery Products', Sea Food Export Journal, Vol. XXX (7):19-23.

Shetty, (1987), Manpower development for marine fisheries, keynote address read at the National Symposium on Research and Development in marine fisheries, Mandapam camp, CMFRI, pp. 16-18, September 1987.

Shiva Vandana and Bandyopadhyaya, J, (1982). Political Economy of Technological Polarisation, Economic and Political Weekly, XVII (45), November 6, p. 1829.

Shrivastava, Jyoti and Ahuja Rajeev. (2002). 'Shrimp Turtle Decision in WTO: Some Implication', Economic and Political Weekly, Vol. XXXVII(33):3445-3454.

Sidarto, A. (1981). 'The strategy of artisanalfisheries development of Indonesia', CMFRI bulletin, 30-A, pp.76-79.

Sivagnanam, Jothi K. and Rajendra C. (1995). 'Blue Revolution in Green Belt', Economic and Political Weekly, Vol. XXX (12):607-608.

Smith, I. R. (1979), A research framework for traditional fisheries. ICLARM Studies and Reviews No.,2. Manila: International Center for Living Aquatic Resources Management, 40 pp.

Smith, I. R. and A. N. Mines (1982).Implications for equity and management. In: I R Smith and A. N. Mines (eds.), Small Scale Fisheries of San Miguel Bay, Philippines: economics Reports of production and marketing ICLARM Technical No. 8. Manila: International Centre for Living Aquatic Resources Management; Institute of Fisheries Development and Research, College of Fisheries, University of the Philippines in the Visayas:and the United Nations University.

Somvanshi,V.S. murty and Rao Vedavyasa.(1996). 'Marine Fishery Resources of India – Present Status and Management Concerns' in Menon N.G. and Pillai C.S.G. (eds.): Marine Biodiversity and Conservation and Management. Cochin: Central Marine Fisheries Research Institute.

Sonak, Sangeeta (2002). 'Coastal Aquaculure in India –A Short Commentry on Failed Expectation', Coastin-A Coastal News Letter, No. 6, Pp.8-11

Sreekrishna, and Biradar R.S. (1991). 'Human Resource Development in Fisheries', CMFRI bulletin, 44, Part Three, Pp. 488-491.

Sridevi, C. (1989). 'The Fisherwomen Financer', Economic and Political Weekly, Vol. XXIV (17):Ws.6-Ws.9

Srinath, Krishna and Jacob Jancy. (1998). 'Behavior Pattern of Fisherwomen in Relation to Fisheries Development Programmes', Proceedings of Workshop on Gainful Employment for women in Fisheries Field, March 7-8, Cochin, Department of Science & Technology,

Sudaraj, P. et. al. (1991). 'Marine Fisheries Development in Tamil Nadu', CMFRI bulletin, 44, Part Three, Pp.583-592.

Sukumar, P. et. al. (1991). 'Training Fisherwomen in Fish Processing', CMFRI bulletin, 44, Part Three, Pp.498-502.

Sundaresan, R. (1981). 'Financing Small Fishermen- A Challenging Task', CMFRI bulletin, 30-A, Pp. 162-164.

Swaminath, M. and Rajendran R. (1991). 'Man Power Needs for Marine Fisheries by 2000AD', CMFRI bulletin, 44, Part Three, Pp.480-488.

The Second Indian Fisheries Forum Proceedings, May 27-31, 1990, Mangalore, India pp. 361-363.

Theesan, Jega G. et. al. (1998). 'Gainful Employment for Fisheries Women in Fish Product Production', Proceedings of All India Workshop on Gainful Employment for Women in Fisheries Field, March 2-8, Cochin, Department of Science & Technology.

Thomas, K. T. (1986). 'Operation Marina and Struggle of Fisher People', Economic and Political Weekly, 18 January, Vol. XXI(3): 105-106.

Vijayakumar, K. (2001). 'Social Audit – An ideal Method for Neutralizing Conflict Situation in Aquaculture Industry', Perspective in Mari culture, The Marine Biological Association of India,

Vishwanath, Susan (2000). 'Workers of the Sea', Seminar, 485, Pp.98-102.

Williams, M.J. and Nandeesha, M.C. (eds). (1998). 'International Symposium on Women in Asian Fisheries', Proceedings of Fifth Asian Forum on Women, November 13, Thailand.

Xavier, F.V.V. (1991). 'Status and Programme of Marine Fisheries Development in the Union Territory of Pondichery', CMFRI bulletin, 44, Part Three, Pp. 577-578.

OFFICIAL REPORTS/DOCUMENTS

'Agriculture Census – Report on main census 1995-96', Goa- Directorate of Planning, Statistics and Evaluation, Panaji- Goa.

'Draft of Ninth Five Year Plan. (1997-2000). and Annual Plan (1997)', Directorate of Planning, Statistics and Evaluation. Panjim. Goa.

'Regional Plan for Goa. 2001 A.D. (Proposal)', Regional Plan Cell, Town and Country Planning Department, Government of Goa. Panjim. Goa.

'Rules of Procedure and Conduct of Business of the Goa Legislative Assembly', Goa Legislative Department, Assembly Hall, Panjim. 1993

DFO (Department of Fisheries and Oceans Canada), (1997), Framework and guidelines for implementing the co-management approval, DFO, Ottawa, 24 pp.

Goa Chamber of Commerce and industry (1990) Industrial and Commercial Directory, Goa. Goa Chamber of Commerce and Industry (1991). '30 Years Of Economic Development', Panjim. Goa.

Goa Foundation, "Fishing File", Mapusa.

Goa Institute of Research, "Fishing File", Panjim.

Goenchea Raponkarancho Ekvott. "Rules of Procedure and conduct of Business of the Goa Legislative Assembly", Goa. 1975-1995 Souvenir.

Government of Goa Gazette series.1 No.15 dated 9-7-1981.

Government of Goa. (2000) The Indian Fisheries Act 897 and The Goa Fisheries Rules. 1981. (Incorporating amendments up to date),

Government of Goa. Daman and Diu Law Department (legal advice) Notification LD/6/34. Government Printing Press - Panjim Goa 162/500-9/1981.

Government of Goa: "The Indian Fisheries Act, 1897" and "The Goa Fisheries Rules, 1981".

Ministry of Fisheries and Agriculture, Republic of Maldives. (1991). 'Status and need of Fisherfolk, Vaatu, Meemu and Faatu Atolls, Maldives', Madras: Bay of Bengal Programme News

Official Gazette-Government of Goa Series 1, No.7. 27 July 2000.

T. R. Thankappan Achari and M. Devidas Menon, (1963). A Report on the Assessment of the Indo-Norwegian Project on the Socio-Economic Conditions of the Fishermen of the Indo-Norwegian Project Area, (NORAD, Oslo)

Unpublished Debates of Goa Legislative Assembly 2000 21-7-00 5.55(1)

PH.D. THESES AND DISSERTATIONS

Bhatta, Ramachandra (1991). 'The Impact of Supply Variation on Fishermen Income', M.Sc. Dissertation. University of Hull, England.

Bhatta, Ramachandra (1994). 'Some Aspects of Marketing of Fish in Karnataka', Ph.D. Thesis submitted to Mangalore University.

John K C (1994). 'Production and Marketing of Fish – Case Studies in Karnataka', Ph.D. Thesis Submitted to Bangalore University, Guide Dr. R. G. Desai.

NEWSPAPER REFERENCES

Gomantak Times, June-July-August 2000.

Gomantak Times, 25 Th July, 2000.

Herald, 28th July, 2000.

Herald, 15th June, 2002.

Herald, 9th July, 2002.

Herald, 5th July, 2002.

Herald, 10th July, 2002.

Herald, 25th July, 2002.

Herald, 31st July, 2002.

Herald, 1st August, 2002.

Herald, 3rd August, 2002

Herald, 7th August, 2002.

Herald, 10th August .2002.

Herald, 11th August. 2002.

Herald, 14th August. 2002.

Herald, 27 Th August. 2002.

Navhind Times June-July-August 2000.

The Hindu, 26th April. 2002.